RAND McNALLY

Historical Atlas and Guide

Rand McNally
Historical Atlas and Guide

Published and printed by Rand McNally
in 1993 in the U.S.A.

Copyright © 1991 by
J.W. Cappelens Forlag A/S, Oslo;
Maps International AB, Stockholm
Liber Kartor AB, Stockholm;
as *Concise Earth History.*
Revised 1993.

Editors and Consultants:
Anders Røhr
Prof. Knut Mykland
Lars Ove Larsson
Tor Åhman

Translation:
Jim Manis, Randy Morse

Drawings in upper margin:
Ellen Bakke

Technical Production:
Boksenteret A/S, Oslo

Production Management:
Bo Gramfors, Siv Eklund
Maps International, Stockholm

Library of Congress Cataloging-in-Publication Data
Rand McNally and Company.
    Historical atlas & guide.
      p. cm.
    Includes index.
    ISBN 0-528-83623-4
    1. Historical geography—Maps.   I. Title.   II. Title: Historical
atlas and guide.
    G1030.R33 1993 <G&M>   93-11581
    911—dc20   CIP
                                      MAP

Preface

Our lives are borrowed from history. Events, decisions, accidents, incidents and sheer coincidence shape history as well as our daily lives.

History is formed by people, through their interactions with one another and nature. Through genius and blind ambition; lust for power and quest for grace; wisdom and folly; love and hate, previous generations have presented us a stage upon which to act out our lives. It is now up to us—each of us—to decide what part we will play.

How will we be remembered? As the final arbiters of planetary peace, or torch bearers passing on the age-old flames of death and destruction?

Will our age be looked back upon as an era of irreversible environmental degradation, or as the time when ecological madness on a global scale was finally brought to a halt?

Will future generations remember the last years of the 20th century as a pivotal—and positive—period for humankind?

Only time—and future historians—can tell. One thing, for now, is certain: moving ahead is always made easier by knowing what has gone before. This book will provide you with basic background. As for future history, it is yours for the making.

Bo Gramfors

Contents

Reconstruction of a dwelling from ca. 20 000 BC.

Mammoth carved in horn. Probably a 15 000-year old handle.

The «Venus» of Willendorf. From Early Palaeolithic times.

Left: The Black Ox. Detail from a 15 000 year old (est.) cave painting from Lascaux in France (map nr. 7, p. 9). The Lascaux cave contains the largest and best preserved man-made images from prehistoric times.

1. THE EARLIEST HUMANS

The Earliest Humans

We don't know anything for certain about mankind's earliest roots – not yet. We can't even say with complete accuracy where these origins are to be found. Nor can we say exactly when the first tools were produced, or when language evolved. These remain among the many unsolved riddles of our common human heritage.

Yet, in spite of the fact that some fundamental questions remain unanswered, year by year we add a bit more to our knowledge of mankind's age, origins and evolution. In Asia, and especially in Africa, many discoveries have been made, in the past as well as in more recent times, that help to shed some light on the haziest chapters in mankind's biological prehistory.

Certain areas of southern and eastern Africa have proved to be especially productive hunting grounds for geological and paleontological «man-hunters». New dating methods have made it possible to date skeletal remains with far greater accuracy than ever before.

While only a few years ago researchers might have had to guess at a number to indicate the absolute age of their finds, they can now order their material sequentially millions

2. AREAS OF EARLY AGRI-CULTURE AND CATTLE-RAISING

Around 8000 BC
Occuring between:
8000-7000 BC   7000-6000 BC
6000-3000 BC.   Numbers in italics give approximate dates

Plowing with wooden plow. From a rock-carving site in Italy.

 Skiers. From a rock-carving site in Karelia.

 Ice Age hunter outside his shelter of mammoth tusks. Reconstruction.

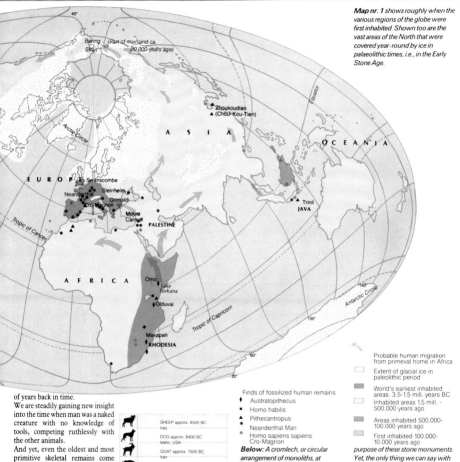

Map nr. 1 *shows roughly when the various regions of the globe were first inhabited. Shown too are the vast areas of the North that were covered year-round by ice in palaeolithic times, i.e., in the Early Stone Age.*

Probable human migration from primeval home in Africa

Extent of glacial ice in paleolithic period

World's earliest inhabited areas. 3.5-1.5 mill. years BC

Inhabited areas 1.5 mill. - 500.000 years ago

Areas inhabited 500.000-100.000 years ago

First inhabited 100.000-10.000 years ago

of years back in time.

We are steadily gaining new insight into the time when man was a naked creature with no knowledge of tools, competing ruthlessly with the other animals.

And yet, even the oldest and most primitive skeletal remains come from the species known as *Homo Erectus* – «upright-walking man». It now seems most likely that Africa was the area on Earth where creatures like us first evolved. It is at any rate here that the fossil remains are most numerous, oldest and best documented.

Anders Hagen, *Cappelens verdenshistorie*, v.1

The drawing at the left *gives an overview of a number of domestic animals, with approximate dates of domestication. Beneath the dates the areas in which they were first domesticated are also given.*

Finds of fossilized human remains
♦ Australopithecus
■ Homo habilis
● Pithecantropus
○ Neanderthal Man
● Homo sapiens sapiens Cro-Magnon

Below: *A cromlech, or circular arrangement of monoliths, at Keswick in Cumberland, Great Britain from the megalithic period (approx. 3000 BC). Several theories have been offered to explain the*

	SHEEP approx. 8500 BC Iraq
	DOG approx. 8400 BC Idaho, USA
	GOAT approx. 7500 BC Iran
	PIG approx. 7000 BC Turkey
	OX/COW approx. 6500 BC Anatolia
	LLAMA approx. 3500 BC Peru
	DONKEY approx. 3000 BC Egypt
	CAMEL approx. 3000 BC Southern part of Soviet Union
	DROMEDARY approx. 3000 Saudi Arabia
	HORSE approx. 3000 BC Ukraine
	CHICKEN approx. 2000 BC Pakistan
	CAT approx. 1600 BC Egypt
	GOOSE approx. 1500 BC Germany
	ALPACA approx. 1500 BC Peru

purpose of these stone monuments. Yet, the only thing we can say with some certainty is that they must have been connected to religious practices.

Reconstructive drawing of a grave from the Halstatt period.

Woman spinning with a spindle. Decoration on clay jar from Hungary.

Summersaulting over a bull. Mural from Knossos.

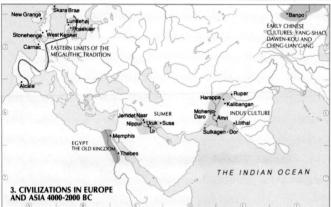

3. CIVILIZATIONS IN EUROPE AND ASIA 4000-2000 BC

4. THE HALSTATT CULTURE

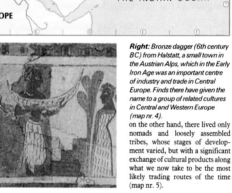

Above: A funeral procession placing sacrificial animals and containers of wine or blood before a priest. From a sarcophagus from Hagia Triada in the Palace of Knossos (see right map). The man leading the procession is carrying a ship, symbolizing the impending cosmic journey.

Right: Bronze dagger (6th century BC) from Halstatt, a small town in the Austrian Alps, which in the Early Iron Age was an important centre of industry and trade in Central Europe. Finds there have given the name to a group of related cultures in Central and Western Europe (map nr. 4).

on the other hand, there lived only nomads and loosely assembled tribes, whose stages of development varied, but with a significant exchange of cultural products along what we now take to be the most likely trading routes of the time (map nr. 5).

Civilization moves westward

From about the year 1500 BC we can trace established, orderly societies in the east, with towns and organized trade. In most of Europe,

5. EUROPE AND THE MIDDLE EAST 1500-1000 BC

- Areas with organized trade, government, towns and villages
- Sparse agricultural settlement
- Nomads
- — Trade routes
- → Major trade routes in Europe

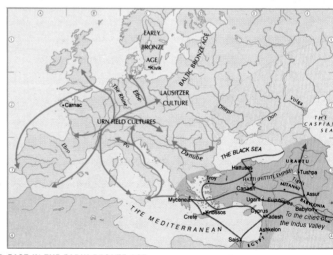

Log house from La Tène period. Reconstructive drawing.

 Celtic warrior on horseback. Relief from 4th century BC.

 Grave monument from Monolithic period. Cornwall, S. England.

7. CAVES FROM PRE-HISTORIC TIMES IN SOUTHERN FRANCE AND SPAIN

6. CELTIC EXPANSION TO 200 BC - LA TÈNE

Before 400 BC

400-200 BC

Celtic civilization

The Celts first settled in the central areas of Europe, but sometime between 400 and 200 BC they began migrating both west and east (map nr. 6). In La Tène by Lake Neuchâtel in Switzerland large numbers of iron weapons and tools from this period have been found.

Above: Detail from a grave complex, West Kennet in southern England (map nr. 3). From the so-called Megalithic Period, a name taken from Greek: megas large, and lithos stone. We can only marvel that people 5 000 years ago, without the benefit of machinery, managed to transport and erect these heavy, roughly hewn stones.

Right: The fortress town of Biskupin on an island in western Poland (map nr. 6). Ca. 30 000 oak timbers have been driven into the ground, to protect the town from the marsh, as well as against enemies. A causeway of oak leads to the mainland. The houses, all built of pine and oak timbers, are thatched with sod. By the main gate is a small market square. At the height of its development, i.e., around 550-400 BC, Biskupin must have had at least 1500 inhabitants (Drawing by Lars Tangedal)

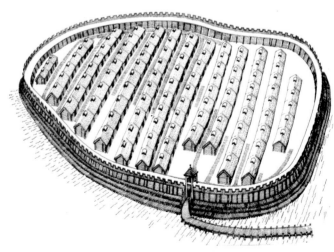

Dancing girl from Mohenjo-Daro.

Axes from the Indus culture.

Hittite warrior from the 14th century BC.

8. MESOPOTAMIA AND THE INDUS REGION

▧ Kingdom of Sargon I 2330 BC
▨ Indus cultures
— Trade routes

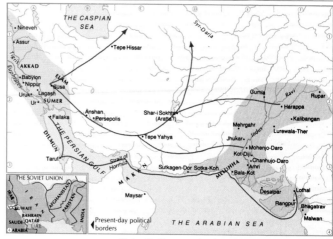

The Indus Valley Civilization

Along the Indus River in present-day Pakistan a unique and highly advanced civilization existed from about 2500–1800 BC (map nr. 8). In the early 1920s extensive excavations were carried out at the sites of two ancient cities, Harappa and Mohenjo-Daro (illust. p. 11). Both here, and at other sites excavated later, archaeologists have uncovered ruins of highly sophisticated cities. Harappa and Mohenjo-Daro are thought to have had around 40 000 inhabitants each.

Below: Soldiers from Kush (Nubia), the area between the 2nd and 3rd cataracts (map nr. 10). Wood carving from the Second Kingdom in Egypt (ca. 2000 BC). The total size of the artifact is 193 × 73 cm, with forty soldiers marching four abreast. It is now in the Cairo Museum.

Warrior King Thutmos III

Because of his many campaigns of conquest, Thutmos, who became Pharaoh in Egypt in 1468 BC, has been called the Napoleon of Antiquity.

During his thirty-year reign, he undertook a total of seventeen campaigns of conquest in Asia. The eighth of these took his troops all the way to the Euphrates.

Under Thutmos's predecessor, Queen Hatshepsut, whose reign had been quite peaceful, the small Asiatic states of Palestine and Syria had formed a coalition, led by the new Hurrittic kingdom of Mitanni. It was this coalition, and the Hittite Empire further to the north, that Thutmos III wanted to bring into line. His soldiers won many battles, and much booty – in the form of slaves, cattle and precious metals – was brought home to Egypt.

However, when Thutmos died at about the age of 55 in 1438 BC, the Mitanni Empire had yet to be subdued.

The map below shows areas attacked by Thutmos III's legions, and counter-attacks by the Hittite and Mitanni empires.

9. THUTMOS III's CAMPAIGNS

Sumerian chariot
from Ur. Ca. 2500
BC.

Mesopotamian warrior
with tunic and
battle axe.

Hammurabi standing
before the sun-god
Shamash.

Left: The excavation of Mohenjo-Daro in the Indus Valley soon revealed that all the houses had been built of very costly fired ceramic brick. Seen here is the same standardized brick (7 × 14 × 28 cm), used in one of the city's wells.

Right: Statue from Uruk in Sumer is from ca. 3000 BC, thought to represent one of the very first Mesopotamian rulers. The figure, eighteen cm high, is worked in alabaster. The eyes are mother-of-pearl and lapis lazuli, inlaid in asphalt.

lished Mesopotamia's first high civilization about 3000 BC, with Uruk as its cultural and administrative centre. The entire area corresponds approximately to present-day Iraq.

It was not, however, until around 1800 BC that the great law-giver, king Hammurabi, unified most of

the area in one kingdom with Babylon as its capital (map nr. 10). Babylon was one of the youngest cities in Mesopotamia, but it became in Hammurabi's time the cultural hub of the entire Near East, a position the city maintained long after it had lost its independence.

The Land Between the Rivers

Mesopotamia is originally a Greek name and means «the land between the rivers», in this case, between the Euphrates and the Tigris. Settlements some 12000 years old have been discovered in this area, and in the south, the Sumerians estab-

Below: Statue of the Egyptian feudal lord Ka-aper, who governed one of the country's in all 42 fiefs or noms. Found in Sakkara, it is one of the most famous artifacts from the Old Kingdom (2665–2155 BC). Now in the Cairo Museum.

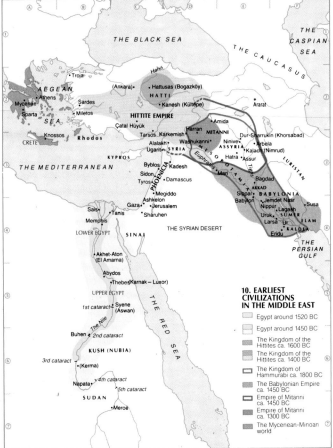

10. EARLIEST CIVILIZATIONS IN THE MIDDLE EAST

- Egypt around 1520 BC
- Egypt around 1450 BC
- The Kingdom of the Hittites ca. 1600 BC
- The Kingdom of the Hittites ca. 1400 BC
- The Kingdom of Hammurabi ca. 1800 BC
- The Babylonian Empire ca. 1450 BC
- Empire of Mitanni ca. 1450 BC
- Empire of Mitanni ca. 1300 BC
- The Mycenean-Minoan world

Tiled dwelling from Ur. Attempted reconstruction.

Temple tower (the Ziggurat) in Ur. 22nd century BC.

Helmet hammered out of 15 carat gold. From Ur in Sumeria.

11. THE KINGDOM OF SARGON I IN THE 24th CENTURY BC

Above: In 2382 BC Sargon I came to power in the city of Kish in Akkad, Mesopotamia. The map shows the boundaries of the kingdom he established during the fifty-six years of his reign.

Left: Warrior from the 8th century BC, i.e., at the time when the Assyrian empire had reached its peak (map nr. 12). The Assyrians, armed with coats of mail, spears and shields, were fearsome warriors.

The Treasures of Ur

The first archaeological studies in Ur in Chaldea were undertaken by J.E. Taylor in 1854. But it wasn't until the end of the 1920s that the famous Treasures of Ur were uncovered during excavations lead by the British archaeologist Sir Leonard Woolley. Several royal tombs were found that had not been looted in ancient times, brimming with jewelry of gold, silver and precious stones, together with weapons and golden vessels. Today the nearly 5 000 year old treasures from Ur are one of the major attractions at the British Museum in London.

12. THE NEAR EAST AROUND 600 BC
- The Kingdom of the Medes
- Cilicia
- Lydian Empire
- The New Babylonian Empire
- — The Assyrian Empire ca. 700 BC

Darius I on lion hunt. Relief from 5th or 6th century BC.

Median princes walking up the stairs to Persepolis.

Scythian warrior on horseback. Statuette from 5th century BC.

Left: Detail from one of the artifacts from Ur, a billy goat, produced in gold leaf, lapis lazuli and mother-of-pearl.

Right: A gilded wagon with a four-horse team carries a Persian king or satrap. Miniature from about 500–200 BC, belonging to the so-called Oxus Treasure. Now in the British Museum, London.

The Persian Empire
559–479 BC

In 559-529 BC Cyrus II of the Achaemenid, becomes king of the Persians, conquers the Median Empire, defeats the Lydian king Croesus, and annexes Lydia and the Greek city-states on the coast of Asia Minor in 546. He conquers Babylon and allows the Jews to return to Palestine. Cyrus II falls against the Scythians in 529.

529–522 King Kambyses II rules the empire and conquers Egypt in 525. He also carries out campaigns in Libya and Nubia.

522–479 Darius I founds Susa and Persepolis and organizes the empire into twenty administrative districts, with standing armies at the Kings disposal. Darius attacks Egypt in 518, advances toward the Indus in 513, crosses the Bosporus in 512 and makes dependencies of Thrace and Macedonia, but is stopped by the Scythians. A rebellion in the Greek city-states in Asia Minor is put down in 500–494. A punitive mission against Greece ends in the destruction of Eretria, but the Persians suffer defeat at the hands of the Athenians under Miltiades at Marathon in 490. In 485 Xerxes I comes into power in the Persian Empire and he puts down revolts in Babylonia and Egypt. Under his campaign against the Greeks 480–479 he defeats the spartans under Leonidas at Termopylene and Athens is burnt down. But the Persian fleet is crushed by Themistokles at the battle of Salamis and the Persian army is routed while withdrawing, following the battle at Plataiai in 479 BC.

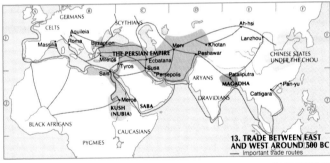

13. TRADE BETWEEN EAST AND WEST AROUND 500 BC
— Important trade routes

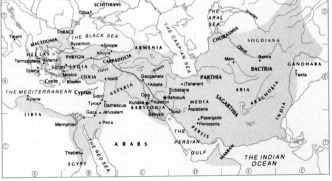

14. THE PERSIAN EMPIRE AROUND 500 BC
Campaigns against Greece 490 BC
Xerxes' attacks Greece 480 BC
The royal highway

Dead man's heart is weighed against the feather of truth. Egyptian papyrus.

Egyptian woman playing a harp. Detail from a grave painting.

Egyptian woman grinding corn with a hand-mill.

15. THE EGYPT OF THE PHARAOHS

▨ Arable land

▧ Desert and steppes

▲ Pyramids

◇ Quarries and mines

▬▬▬ Southernmost border 2665-2155 BC (The Old Kingdom)

▬▬▬ Southernmost border 2061-1650 BC (Middle Kingdom)

— Caravan routes

Left: *One of Egypt's best known Pharaohs is Tutankhamen, (1346–1337 BC). His fame does not rest so much with his political deeds as with the sensation created by the opening of his tomb in the 1920s. Among the many treasures was the king's throne, carved in wood, covered in gold and silver and inlaid with gems, faience and glass. In the detail of the back of the throne shown here, Tutankhamen himself is seen in all his splendour.*

Below: *A detail from an Egyptian tomb painting from the period of the New Kingdom – Woman with duckling.*

Egypt 3050–1438 BC

About 3050 BC begins a period of national consolidation. The two centres of power, Lower Egypt in the Nile delta and Upper Egypt in the south, are united under one ruler.

The Old Kingdom

Around 2630 BC the royal architect Imhotep builds a stair-step pyramid, the Stairway to Heaven, at Sakkara. Power is centred around the Pharaoh and his residence in Memphis, the country's only capital city.

2575–2550 Under king Cheops kingship develops into absolute monarchy. Construction of the great pyramid of Cheops is begun.
2540–2515 In king Chephren's day the Sphinx is carved out of sandstone.
2480–2154 A new dynasty, the «Sons of Ra», has come to power. Majestic temples are built in honour of the sun god, Ra. Art and culture blossom. With time highly placed civil servants become royal princes, with large, hereditary estates. Rivalries develop, leading to internal squabbling. Finally, royal power is weakened and the

An unconventional
form of coitus.
Drawing from an
Egyptian papyrus.

Cross section of
Cheops pyramid with
the king's chamber
in the centre.

Egyptian Ibis from
the 3rd century BC.

16. THE PYRAMIDS AT GIZA
Smaller tombs around the pyramids
are not marked

Right: The pyramids at Giza. In the foreground queens' pyramids in front of the pyramid of Mycerinus, followed by the pyramid of Chephren, and furthest back, the grandest of them all, the great pyramid of Cheops. Together they cover an area of about 200 acres, and the distance in a straight line from the nearest corner of the pyramid of Mycerinus to the farthest corner of Cheops is approximately 1.2 km. We can get some idea of the size by observing the two riders. The Cheops pyramid was 230 m at the base.

country falls into a state of chaos. 2154-2000 All against one, and one against all. The royal princes exploit the peasants mercilessly. The tombs of the kings are plundered. The country is divided into small fiefdoms.

The Middle Kingdom
Ca. 2000-1970 BC Amenemhet I unites the kingdom once again. A great fortification, «the Walls of the Kings», is built across the Isthmus of Suez. Under Amenemhet and Sesostris III, Egypt experiences a golden age of literature, science, art and architecture.
Ca. 1860 The «Land of Gold», Nubia, is conquered to the Second Cataract. The End of the fiefdoms.
1650-1544 BC The age of the Hyksos. Asiatic peoples migrate into the Kingdom and occupy large areas of land. A Semitic dynasty takes power in the Delta, while the southern parts remain in the hands of the Egyptians.

The New Kingdom
Ca. 1540 BC the Asiatic invaders are driven out of Egypt, the Nubians are defeated, and Egypt is reunited. Thebes again becomes the country's capital, and Amon-Ra the chief deity.
1468-1438 BC Under Thutmos

III Egyptian conquest was carried to the Euphrates, including the northern Sudan, Phoenicia and Syria (map nr. 10, p. 11).

Israel ca. 1030-931 BC
Approx. 1030-1015 BC Israel's first king, Saul, battles the Philistines who are invading the country. He is successful at first, but falls in

17. ISRAEL UNDER DAVID AND SOLOMON (CA. 1005-925 BC)
The Kingdom of Israel

battle in 1015.
Ca. 1015-972 David succeeds Saul as king. He defeats the Philistines, conquers Jerusalem, which becomes his governmental seat, and builds a tabernacle to house the «Ark of the Covenant», for the stone tablets with the Ten Commandments written by God.
Ca. 972-931 Under king Solomon the kingdom has its period of greatest achievement, blossoming both intellectually and materially.
931 The kingdom is divided after

Solomon's death. The rebel leader Jeroboam becomes king of Israel, the area inhabited by the northern tribes, while Solomon's son, Rehoboam becomes king of the southern tribes in Judah.

18. ISRAEL AND JUDAH CA. 860 BC
The Kingdom of Israel
The Kingdom of Judah
Assyria's southern border 721 BC

Female athlete from Sparta. Contemporary bronze statuette.

Reconstructive drawing of a bourgeois home in Athens.

Phoenician ships, loaded with logs. Relief from the 8th century BC.

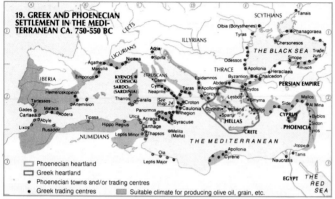

Greece ca. 750–449 BC

750–650 Overpopulation and the resulting food shortage leads to emigration from mainland Greece. Greek men are forced to seek out new land. Agricultural settlements are established in Sicily, southern Italy, southern France and northern Africa (map nr. 19).

The 6th Century BC Aristocratic constitutions are overturned by revolutions in several Greek states, where tyrants seize power. The Age of Tyrants is interspersed with periods that see democratic constitutions evolve.

Ca. 600 The first method of striking coins is invented. This leads to an increase in trading activities.

594 Solon proposes a new constitution for Athens.

Ca. 570 Sparta is one of the mightiest military powers of the age.

552–527 The tyrant Peisistratos lays the groundwork for an Athenian age of greatness.

546–449 The cities in Asia Minor submit to Persian rule, and Persia attempts to conquer Greece itself (see p. 13). The Persian Wars begin with rebellions in the Greek cities in Asia Minor in 500, and the Persians are not stopped before they have reached Marathon (490) and Salamis (480). These victories make Athens a great-power, and Sparta's bitter rival.

Above: A votive shield from Greece (6th century BC) depicts an offering to the gods, as gifts and various trappings are carried in a procession to the alter. A young boy is ready with the offer while the alter is sprinkled with consecrated wine.

The Phoenicians

were known for their prowess as seafarers and traders. The heart of this activity, the Phoenician homeland, was present-day Lebanon (map nr. 19). Tyros and Sidon became the major trading centres in the ancient world.

The Phoenicians traded throughout the Mediterranean. In the period 750–550 there was also extensive emigration to new settlements in this area. Carthage and Utica in present-day Tunisia were among the first to be established.

20. CENTRAL GREECE IN ANTIQUITY

- Dorians
- Ionians
- Aeolians
- Arcadians
- Northwestern Greeks
- — Important road

Loom with counter-weights from classical Greece.

The Acropolis of Athens, built in the 5th century BC.

Pericles, Athens' leader during her period of greatness in the 5th cent. BC.

21. THE PELOPONNESIAN WAR 431-404 BC

▨ Athens, The Delian League and its allies
▨ Sparta and her allies
☐ Neutral states
⟶ Main campaigns by Athens and the Delian League
⟶ Main campaigns by Sparta and her allies
★ Important battle sites with dates

The Delian League

In 478 BC a maritime confederation was formed between Athens and other Greek city-states, primarily a defense league against the Persian Empire, under the military leadership of Athens. The League's headquarters was in Athens as well, but the war chest was maintained in the Temple of Apollo on the island of Delos.

The allied states were to provide either warships or money. Since most parties preferred the latter option, the League's fighting fleet and the Athenian navy soon became one and the same force.

During the Peloponnesian War (see below) it became a major objective for Sparta to crush the Delian League, and with Sparta's victory in 404 BC the League was dissolved.

The Peloponnesian War

The reign of the great statesman and strategist Pericles (ca. 500–429 BC), Athens' age of greatness. Not only did the city become a centre of international trade, but the Athenians were also busy building up an empire in the Mediterranean. This was the reason Sparta went to war with Athens in 431.

It is called the Peloponnesian War, and did not end until an alliance of Spartans and Persians crushed the main body of the Athenian navy at Aegospotami in 405. Only eight vessels, under the strategist Conon, managed to make their way to Cyprus. One year later Athens had to capitulate.

Right: The Greek dramatist Sophocles (496–406 BC) was influenced by myth and legend, but his topics and his characters made the plays contemporary dramas that had important things to say about the politics of his day. F.ex. in his «Oedipus Rex», which is not only Sophocles' masterpiece, but that of ancient drama on the whole. Here a modern production of the play at the Herodes Atticus Theatre at the Acropolis (map nr. 23, p. 19).

Discus thrower.
Greek marble statue
from ca. 450 BC.

Greek distance
runners.

Olympic sprinter ca.
500 BC.

Above: The ruins of the Pronaia temple in Delphi, which, like the temple of Apollo, was destroyed during the great earthquake in 373 BC. Three years later, however, it was rebuilt. It was at the cult centre of Delphi that the sibyl Pythia delivered the judgments of the Oracle.

Below left: Greek sprinters as depicted on a victory vase from around 470 BC. *Right:* Charioteer from Delphi. Bronze statue from around 470 BC. Both in Olympia and at the Pythian Games in Delphi, the chariot races were a highpoint of the games.

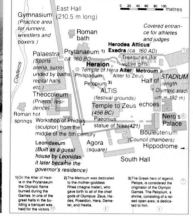

The Olympic Games

The first athletic contests in Greece probably took place in Olympia in 776 BC, or even earlier.

They were held in the summer every four years, and lasted five days. On the first day participants made offerings, prayed to the gods and took the Olympic oath. The following day there were chariot races in the Hippodrome, then the pentathlon in the stadium, with competition in the discus, long jump, javelin, sprints and wrestling. The third day was the day of the full moon. This day began with religious rituals in the morning, and sprints, boxing and wrestling for juniors in the afternoon.

On the fourth day there were two races, one of one stadium length, and one covering two lengths (1 stadium length = 92 m), and a distance run of twenty-seven stadium lengths (c. 5 kms). In the afternoon contestants competed in boxing, wrestling and **pankration**, no-holds-barred combat. The final event of the games was a race, twice the length of the stadium, in full armour.

The fifth day was the awards ceremony.

22. OLYMPIA

Buildings from Classical period (5th century BC)

Constructed in the Hellenic period

From the time of Imperial Rome

The earliest known Olympic games were held in 776 BC, the last in 394 AD

It was during an excavation in Olympia that French archaeologist and pedagogue, Baron Pierre de Coubertin (1863–1937), first had the idea of reviving the Games. The first modern games were held in Greece in the summer of 1896. The first Winter Games took place in Chamonix in 1924.

Grecian temple in the Dorian style. Ca. 450 BC.

Ionian column and capital arch from the Acropolis 430s BC.

Greek amphitheatre from around 350 BC.

Above: *The world around 500 BC, according to the Greek philosopher and scientist Hekataios. Libya (Africa) has been run together with Asia, thus becoming about the same size as Europe. Hekataios produced also a «Description of the World», containing hundreds of place names.*

Left: *One of the colonnades in the Parthenon of the Acropolis.*

23. THE ACROPOLIS OF ATHENS

▨ Buildings from Classical period (5th century BC)

▨ Constructed in the Hellenic period

▨ From the time of Imperial Rome

The Acropolis of Athens

Construction of the Acropolis in Athens in the classical period was begun under Pericles in 447 BC, and the first building completed was the Parthenon according to drawings by the architects Iktinos and Kalikrates. The temple, erected in honour of the goddess Athena, the city's guardian deity, was built in the Dorian style and dedicated in

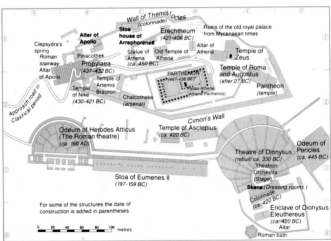

For some of the structures the date of construction is added in parentheses

0 20 40 60 80 100 metres

Left: *During performances at the Dionysian and other Greek theatres, music was an important element. Seen here is an actor with a tambourine, made of a round wooden frame with hide stretched across one side. Bits of metal sound when the instrument is struck or shaken.*

438 BC. A short time later (437–432 BC) the pompous complex of arched entryways, the Propylaea, was built in both the Dorian and Ionic styles. Pericles' Odium, which served as a concert hall, was

added around 445 BC. It was situated close to the Theatre of Dionysus, rebuilt around 330 BC.

From 421 to 406 a temple – Erechtheum – was built north of the Parthenon, and dedicated to Athena, Erechtheus and Poseidon. Construction was begun in order to occupy workers in a period of great unemployment. Pericles commissioned the sculptor Peidias to decorate the Parthenon and the rest of the Acropolis. Among other things, he created the statue Pallas Athene (Athene Parthenos) in gold and ebony. Peidias had his own

studio in Olympia (map nr. 22, p. 18), with assistants to help him complete these enormous projects. In the Hellenistic period (323–30 B.C) a temple of Zeus and the Peristyle of Eumenes II were built, the latter as a kind of foyer to the Dionysian Theatre. During the time of Imperial Rome too, additions were made to the Acropolis – a Roman theatre and, typically, a Roman bath.

The Acropolis of Athens – the city's fortress – is one of Athens' greatest attractions, even today.

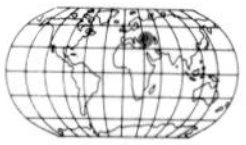

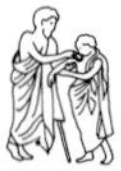

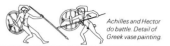

A sick Greek gets treatment ca. 400 BC.

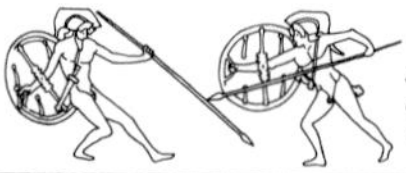

Achilles and Hector do battle. Detail of Greek vase painting.

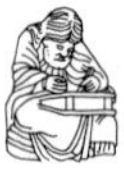

Pythagoras at his desk. Detail from the cathedral in Chartres.

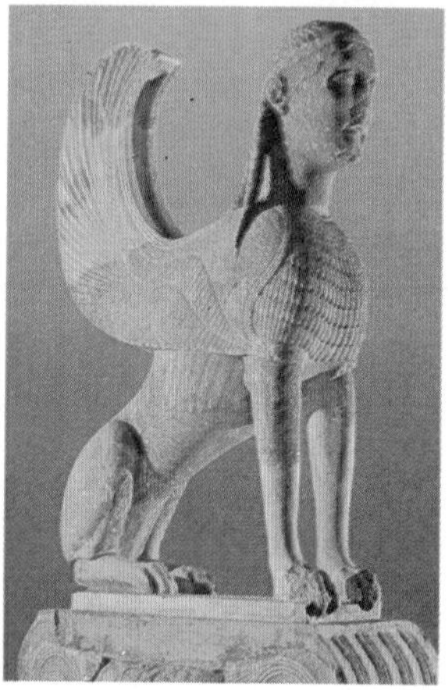

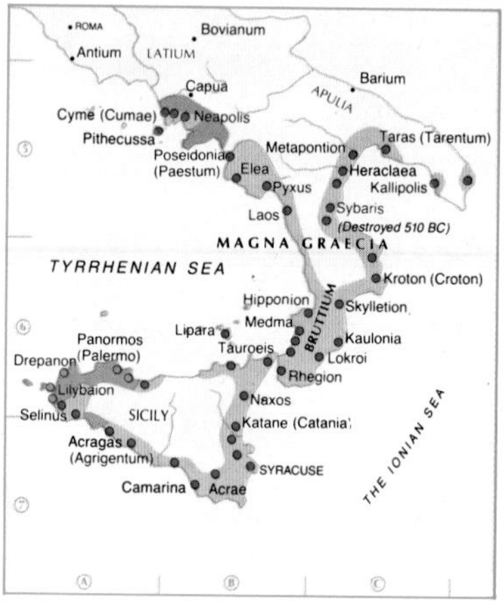

24. GREEK AND CARTHAGINIAN COLONIZATION IN SICILY AND SOUTHERN ITALY. MAGNA GRAECIA (GREATER GREECE)

Areas primarily under:

- Greek influence
- Carthaginian influence
- Etruscan influence
- Greek colony
- Carthaginian colony
- Other towns

26. THE EMPIRE OF ALEXANDER THE GREAT 323 BC

- The empire's greatest expansion
- States dependent of Alexander
- Alexander's campaigns 334-323 B
- Nearkhos's expedition 325 BC
- Alexander's important battles
- Cities founded by Alexander

Above: *Some of the inhabitants of Naxos in the Aegean set out and founded a colony in Sicily. Others who remained at home had from time to time to ask the Oracle in Delphi for advice. This sphinx, a little over two metres high, is a gift from the inhabitants of Naxos to Delphi, perhaps in thanks for a beneficent answer.*

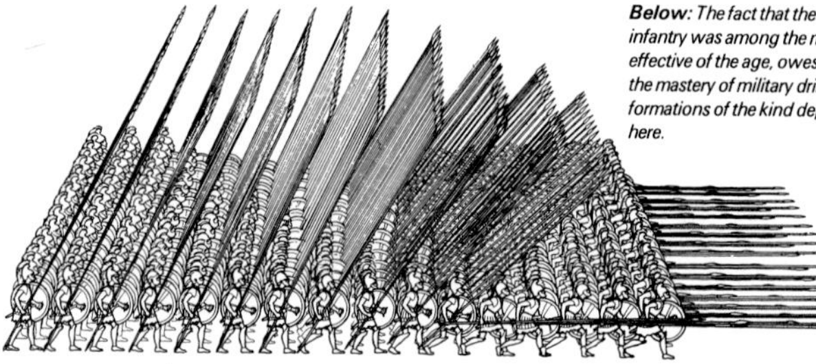

Below: *The fact that the Greek infantry was among the most effective of the age, owes much to the mastery of military drill and formations of the kind depicted here.*

Magna Graecia – Greater Greece

During the period of Greek expansion into the western Mediterranean about 750–550 many emigrants settled and remained in southern Italy and on the island of Sicily (map nr. 24). These settlements quickly grew powerful, and cities like Cyme, Taras, Syracuse and Sybaris – known in particular for their affluence and luxury – became wealthier than those cities in Greece that had given birth to them. The temples in Poseidonia, Selinus and Acragas were both larger and more magnificent than anything in the old country.

This was the homeland of the mathematicians Pythagoras and Archimedes and was considered one of antiquity's most flourishing civilizations.

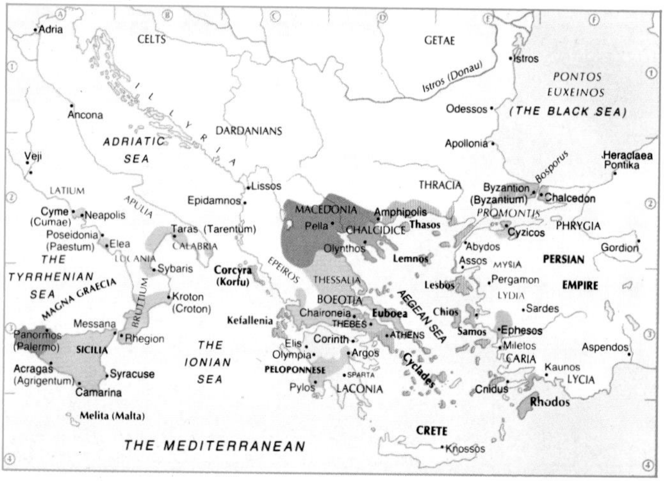

25. THE GREEK WORLD 362 BC

- Athens and the Second Attic Maritime League
- Sparta and her allies
- Boeothia and its allies
- Carthaginian territory

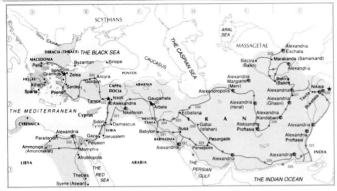

Alexander the Great

When Philip II had fallen at the hands of an assassin, the twenty-year-old Alexander – the son of Philip's fourth wife, Olympias – became king of the Macedonians. The Greeks hoped this would mean an end to Macedonian hegemony, but it soon became evident that «the pup», as he was called, was in fact a young «lion», and after a time the Greeks accepted their new ruler. Only two years after his ascension to the throne, in 334, Alexander led an army of 35 000 men in a campaign of vengeance against the

Below: One who learned much from the geometry of Euclid was Eratosthenes from Cyrene (ca. 275–194 BC), who used Euclidean theories in measuring distances when developing his world map. The axis of the world passed through the Pillars of Hercules (Gibraltar), Rhodes and Issos, and along the Taurus Mts. to the Himalayas.

Right: In late December the Greeks honoured the god of wine and fertility, Dionysus. It was believed that by working themselves into a state of ecstasy, members of the cult could be possessed by the god, thereby sharing in his divinity. It was in particular women who took part in these orgiastic rituals during the feasts of Dionysus. From a vase, ca. 380 BC.

Persian Empire (map nr. 26). The first battle took place near the river Granikos the same year, and here, for the first time, the Greek riders were victorious against the Persian cavalry. In 333 Alexander defeated the Persian king Darius III at Isos, and following the assassination of the Persian king two years later, Alexander laid claim to his throne. He married the Bactrian princess Roxane and made Babylon the seat of his empire. In 323 the 33-year old king died of fever during pre-parations for new expeditions, this time into Sicily, Italy and Africa.

Above: This statue of a horseman depicts the Macedonian king Alexander the Great (356–323 BC). The horse is no doubt the famous warhorse Bukéfalos, which then prince Alexander managed to tame after all others had given up trying. Bukéfalos served his master right up until it was killed during a battle in India in 326 BC. In honour of his horse, Alexander named a city near the site of the battle Bukéfala (map nr. 26). The bronze statue was cast during the Roman period, but is likely a copy of a much older original.

Below: This enlargement of a portrait on a gold coin shows Alexander the Great's father, Philip II of Macedonia (382–336 BC). He ascended the throne in 359 and was a ruthless but very able ruler. He gained control over the gold mines in Thrace and raised a professional army that became the most effective in the world. Philip II united the Greeks in one kingdom.

Etruscan wagon. From grave painting at Volterra.

Etruscan wrestling match. Mural in Tarquinii from about 530 BC.

She-wolf nurses the founders of Rome, Romulus and Remus, at the Capitolium.

The Punic Wars

The Carthaginians – or *punici* – as the Romans called them, had possessions in the western part of Sicily, but by an ancient treaty, the Strait of Messina was to form a boundary between Roman and the Carthaginian spheres of interest. When Carthage carried out naval exercises of Tarentum in Calabria, many in southern Italy saw an excuse to take the rich island of Sicily. The Senate resisted for some time, but in 264 **The First Punic War** broke out.

After 23 years of war, both parties were exhausted, but Roman emerged victorious, and Sicily became a Roman province.

In 218 **The Second Punic War** erupted, and the Carthaginians' famous general, Hannibal, led an army of 40 000 men from Spain, over the Alps and down the entire length of Italy. Yet, in 201 Carthage had to sue for peace.

In 149 **the Third Punic War** began, ending three years later when North Africa too became a Roman province (maps nrs. 28 and 32, p. 24).

27. THE APENNINES (ITALIAN) PENINSULA AROUND 300 BC

- Celtic from ca. 400 BC
- Celtic advances
- Etruscan territory
- Romans and their allies
- Samnitic League
- Greek influence
- Carthaginian territory
- Important road

Lacus Lemanus)((St. Gotthard)
GALLIA CISALPINA
Lacus Verbanus (Lago Maggiore) / *Lacus Larius (Lago di Como)*
VENETIA
Comum
Mediolanum (Milano) Verona *Lacus Benacus (Lago di Garda)* Aquileia Tergeste
Verceliae Augusta Taurinorum (Torino) Cremona Mantua Patavium (Padova)
Placentia Parma Adria (Hadria) Pola Senia
Genua Mutina (Modena)
Vada Sabatia Bononia (Bologna) Ravenna
LIGURIA AEMILIA
Luna Ariminum (Rimini)
Nicaea (Nice) Luca Faesulae (Fiesole) Sena Gallica
Pisae Florentia Sentinum Ancona
MARE LIGUSTICUM (THE LIGURIAN SEA) Arretium UMBRIA Tragurium ILLYRICUM
Lacus Trasimenus Camerinum Pharus
Populonium Volsinii Asculum MARE ADRIATICUM (THE ADRIATIC SEA)
Ilva (Elba) Telamon PICENUM Epidauros
CORSICA Cosa Vulci Veii Corfinium
Alalia (Aleria) Tarquinii Caere (Cerveteri) ROMA Reate SAMNIUM
LATIUM Praeneste Arpi
Ostia Alba Longa Cannae
Tarracina (Anxur) Capua Malventum (Beneventum) Barium
Cumae Herculaneum APULIA Via Appia Brundisium
Turris Libisonis Olbia Neapolis Vesuv
Pompeii Salernum Metapontum Tarentum
Mercurii MARE TYRRHENUM (THE TYRRHENIAN SEA) Paestum LUCANIA Heraclea (Herakleia) CALABRIA
SARDINIA Velia (Elea)
Neapolis Laus Thurii
Carales MAGNA GRAECIA
Nora Croton (Kroton)
Lipareae Insulae (Lipari Islands) BRUTTIUM
Aegates Insulae (Egati Islands) Mylae Messana (Zankle)
Hippo Diarrhytus Panormus (Palermo) Rhegium
Lilybaeum (Marsala) Himera Tauromenium
Selinus SICILIA Etna
Utica Agrigentum (Akragas) Catania (Katane)
Hippo Regius Gela Piazza Armerina Syracusae
Carthage Cossura (Pantelleria) Camarina
Tunes Neapolis
Zama Hadrumetum Melita (Malta)
Leptis Minor
MARE INTERNUM (THE MEDITERRANEAN) MARE IONIUM (THE IONIAN SEA)

Below: The fertility god Demeter (and an armed servant), who, says the Sicilian poet Theocritus (ca. 300–260 BC), is carrying «poppys red, and sheaves of flowering grain in his generous hands». From an Etruscan tomb painting from the 5th century BC, found in Caere.

The Roman Empire 753 – ca. 300 BC

Monarchy
753 According to legend, Romulus and Remus found Rome.
750–575 Latins, Sabines and Etruscans form the *city-state* of Rome, ruled by kings, a senate, clerics and a popular assembly.
Ca. 550 The state is divided into administrative districts for the collecting of taxes and military organization. The society is further divided, according to class: patricians and plebeians.

Republic
The Early Rep. ca. 510–300 BC
Ca. 510 The king is expelled. Executive power is granted to elected officials who serve one year terms, governing in tandem so that *one* man cannot usurp power. Official posts and admission to the Senate are reserved for patricians.
494 Plebeians threaten to leave Rome; patricians forced to accept plebian civil servants and a plebian assembly. The plebeians have won their first victory in the struggle between the estates.
Ca. 493 Rome forms league with the Latin city-states, based on

principles of equality.
451–450 The plebeians win greater protection under the law with the introduction of the 'law of the twelve tablets'.
396 Veii, the most powerful of the Etruscan cities, is conquered. Rome's territory is nearly doubled. Way open to the north.
Ca. 390 Gauls burn Rome.
366–300 Plebeians gain access to patrician civil posts.
340–338 Rome wins a war against alliance of Latin states and Campania. Latins gain all the rights of Roman citizenship – the other defeated allies only half of these.

Etruscan warrior with helmet and sword.

Gallic armor in hammered bronze from time of Caesar.

Roman warrior from time of the Republic. Relief from 4th century BC.

28. THE PUNIC WARS

→ Roman campaigns 218-201 BC
→ Hannibal's campaigns 218-203 BC
--→ Hasdrubal's campaign 208-207 BC

▦ Roman 264 BC
▦ Expansion to 218 BC
▦ Expansion westward 218-121 BC
▭ Carthaginian 264 BC
▢ Carthaginian expansion 238-218 BC
▦ Controlled by Massilia

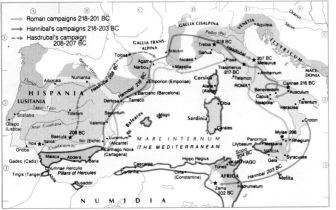

Below: It was against soldiers like those shown here that the Romans had to fight in neighbouring states. Their armour was apparently not much different than that of the Romans themselves. In fact many modern historians point to the fact of Rome's superior numbers as the most likely reason for her success against opponents on the Italian peninsula.

Below: Two stages in the battle of Cannae, where the Roman consul Terrentius Varro engaged the Carthaginian commander Hannibal – during the Second Punic War – in august, 216 BC. Hannibal routed the Romans.

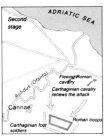

◄ 29. THE BATTLE OF CANNAE 216 BC

Hannibal's divisions: 1 and 5: Cavalry divisions. 2 and 4: African infantry. 3: Spanish/Gallic infantry. 6: Lightly armed Carthaginians.

Varro's Roman divisions: 7 and 9: Infantry. 8 and 10: Cavalry divisions.

30. ASIA MINOR 188 BC ► THE KINGDOM OF PERGAMON

▦ Pergamon 218 BC
▦ Pergamon 218-188 BC
▦ Pergamon 188 BC
▢ The Seleucid Kingdom
▦ Free Greek cities

Roman wagon. The men rode. Woman, children and old people travelled by wagon.

Houses in an Italian city. 1st century AD. Detail from a relief.

Roman carpenter at work. Detail from mural in Pompeii.

Above: *A very famous orator of the Roman period, Marcus Tullius Cicero (106–43 BC). Many of Cicero's orations, letters and philosophical treatises are preserved. Together they give an impression of a lively personality, with a sense of wit and rhythmic diction.*

31. ROME, CARTHAGE AND THE GREEK WORLD CA. 200 BC

- Rome ⎱ Following the Second
- Carthage ⎰ Punic War (201 BC)
- Kingdom of the Seleucidians
- Minor Greek states
- Free Greek states
- Other Greek states
- Egypt and Egyptian possessions
- — Important trade route

For the Kdm. of Pergamon and other nations in Asia Minor, see also map nr. 30, p. 23

32. THE ROMAN EMPIRE AT THE DEATH OF CAESAR 44 BC

- Roman provinces
- Independent states

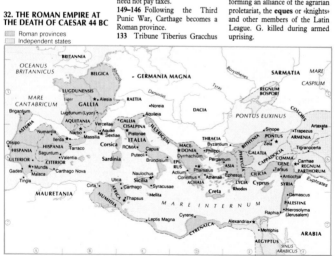

The Roman Empire Ca. 200–44 BC

The Late Republic

200–197 BC War with Macedonia to «liberate» the Greek city-states.
171–168 Macedonia becomes a Roman province. Booty and tribute such that the Romans themselves need not pay taxes.
149–146 Following the Third Punic War, Carthage becomes a Roman province.
133 Tribune Tiberius Gracchus recommends giving state lands to peasants. Because this would make Gracchus a leader of the small farmers, the **optimi**, the party of the nobles, are against the proposal. Gracchus is murdered.
123 Gaius, brother of Tiberius Gracchus, attempts unsuccessfully to push through agrarian reform by forming an alliance of the agrarian proletariat, the **eques** or «knights» and other members of the Latin League. G. killed during armed uprising.

113–101 Cimbrians and Teutons, threaten Rome. Consul Gaius Marius drafts proletariat creating a loyal professional army, whose members expect to be rewarded following each campaign.
88 Mithradates VI of Pontus has 80 000 Romans murdered in the cities in Asia Minor. Senate gives command of the army to Sulla; popular assembly withdraws this post and gives it to Marius. Sulla occupies Rome, and his enemies flee. After Sulla's departure, Marius again takes power.
82–79 Sulla returns after having defeated Mithradates. Marius and his supporters (**populares**) defeated. Sulla becomes dictator, reinstates the power of the Senate.
73–71 Thracian gladiator Spartacus leads a slave revolt that spreads across all of Italy; put down by Crassus.
60 Pompey, Crassus and Caesar form the first **triumvirate**, and so control both the Senate and the Popular Assembly.
58–51 Caesar conquers Gaul. He establishes an army with loyal soldiers.
49 Caesar crosses the Rubicon into Italy. War with the Senate and Pompey, ends with Caesars victory at Farsalos. Pompey flees to Egypt.
48 Pompey is murdered. Caesar makes Cleopatra queen of Egypt. Caesar continues his campaigns in Asia Minor, Africa and Spain.

A Roman catapult. Reconstructive drawing.

Bronze statue of a Roman orator from about 90 BC.

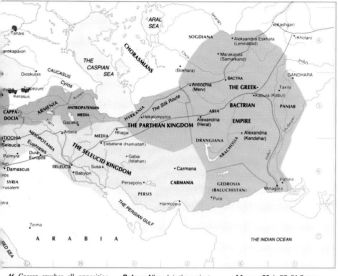

46 Caesar crushes all opposition and becomes dictator for life. He introduces the Julian Calendar.
44 Caesar is murdered by conspiratorial senators.

Below: All roads in the ancient world led to Rome, and all remain today. Here we see one of the most beautiful – the road from Ostia, Rome's seaport.

Map nr. 33: In 58–51 Caesar subjugates Gaul. The population is Romanized as Latin becomes the language of the people. In 55/54 Caesar invades Britain.

Above: This statuette of a Roman legionnaire shows, among other things, that the legionnaires wore coats of mail, constructed of overlapping bands of metal.

33. GAUL AND BRITAIN
- Roman provinces
- Conquered by Caesar 58-51 BC
- Controlled by Massilia
- Caesar's invasion of Britain 55 and 54 BC
- Roman province 43-71 AD

Vallum Antonini
N. border of the Empire 145-181

Vallum Hadriani
N. border 122-367

Eburacum (York) BRITANNIA
Lindum
Deva (Chester)
Viroconium · Ratae (Leicester)
Glevum (Gloucester)
Camulodonum (Colchester)
Aquae Sulis (Bath) Londinium (London)
Noviomagus Aduatuca
GALLIA Castellus Caesaris
OCEANUS BRITANNICUS · Camaracum
Rotomagus (Rouen)
Durocortorum (Reims)
Lutetia (Paris) BELGICA
GALLIA Agedincum
Cenabum (Orleans) CELTICA · Alesia
Portus Namnetum (Nantes) Bibracte
Avaricum (Bourges)
Limonum (Poitiers)
Gergovia Luggo- num (Lyon)
Burdigala (Bordeaux) Genua
AQUITANIA Nemausus NARBONENSIS
Nemes Aqua Sextiae
Tolosa Arelate Forum Julii
GALLIA Narbo Massilia

FRISIANS

GERMANIA

THE ALPS

Roman charioteer with four-horse team. Detail from a mosaic.

The apostle Peter with the Christian cross. 4th century statuette.

Lamp and scale, used in Roman empire.

34. INDUSTRY AND TRADE IN THE ROMAN EMPIRE

- The Empire at the death of Trajan 117 BC
- — Important trade route
- --- Important sea route

SIGNIFICANT TRADE IN AND/OR SOURCES OF:
- ■ Gold ● Wine ■ Horses
- ▲ Silver ▲ Olive oil ● Slaves
- ● Copper ● Grain
- ■ Iron ■ Wool and woolen goods
- ▮ Glass ▮ Linen
- ▼ Pottery ▽ Silk

35. PALESTINE IN THE TIME OF CHRIST

- The Roman province of Syria from 64/63 BC, incl. free states and cities under the Syrian governor
- Kgd. of Judea under the vassal prince Herod the Great 40-4 BC

- Galilee - Perea, Herod Antipas governed 4 BC-33 AD
- Governed by tetrarch Philipus 4 BC - 33 AD
- Samaria - Judea, governed by procurator Pontius Pilate 26-36 AD

Palestine at the Time of Christ

Jesus was born in Bethlehem in Judah, which by this time had become an independent Roman district, separate from Galilee, Samaria and Perea. However, he grew up in Galilee, which together with Perea was governed by the Roman **tetrarch** Herod Antipas. When Jesus was crucified in Jerusalem, Pontius Pilate was «procurator», i.e., Roman governor, in Judea.

Below: A fragment of the gladiator mosaic in the Villa Borghese in Rome. The first known duels between gladiators in Rome took place in 264 BC, and after a time the barbaric custom became a popular diversion. Even Augustus brags that he has ordered 5000 pairs of gladiators into life-or-death combat. The gladiators were often prisoners of war, or violent criminals who could serve out their punishment by performing in the arena for a specified number of years.

Wine press in use. Detail from mosaic in Pompeii.

Roman street musician. Detail from mosaic in Pompeii.

Grain sacks are carried aboard a Roman cargo ship.

36. ROME IN THE TIME OF CAESAR

The four regions of Rome from about 550 BC:

- Collina
- Esquilina
- Palatina
- Suburana (Sucusana)
- Servian Wall, named for Servius Tullius (ca. 550 BC), built ca. 350 BC
- Public buildings

0   500   1000
meter

Above: The Colosseum in Rome as it stands today with the Amphitheatre. The building takes its name from the colossal statue of emperor Nero, Colossus Neronis, which stood nearby. The Amphitheatre, which held an audience of about 50 000, was used primarily for combat, whether it be pairs of gladiators or pairs of wild beasts. It was dedicated in the year 80 BC, one year after Vespasian's death. The length axis of the building is 188 m and the cross axis is 156 m.

37. THE IMPERIAL MARKETS (FORA) IN ROME

Buildings and other structures from:

- Time before Caesar
- Time of Caesar, to 44 BC
- Augustus (43 BC - 14 AD)
- Tiberius - Vespasian (14 - 79 AD)
- Nerva - Trajan (96 - 117)
- Adrian - Antonius Pius (117 - 161)
- Constantine (306 - 337)
- Other periods

The City of Rome

South of the Tiber river lay the area known as Latium. North of the Tiber was Etruria (map nr. 27, p. 22). Commerce between the two regions was carried on via an island in the river (map nr. 36), and east of here it was – on the Pallatium, one of Rome's seven large hills – that the seeds of what was to become the city of Rome first germinated in the 14th century BC. According to legend, however, it was the twin sons of the war god Mars, Romulus and Remus, who founded the city in 753 BC. We can

say with certainty that, beginning in about the year 550, Rome was a growing urban area. And the site on which the Forum Romanum stands today was an important centre even then. Several temples were built, of which the temple of Jupiter, dedicated in 509 BC, was the most magnificent.

In the time of Caesar, the Forum was greatly expanded. He constructed, among other things, the first Imperial market, the Forum of Caesar (Forum Julium). Later several new fora were established, one after the other: Augustus' Forum, Vespasian's Forum (Forum Pacis), Neva's (Domitian's) Forum and Trajan's Forum.

Not only did Rome extend its geographical boundaries to the

fullest under Trajan (98–117 AD), but also became the radiant centre of the ancient world. The inhabitants – over a million – could take pleasure in well furnished baths, libraries and athletic facilities. They could divert themselves with theatre, chariot races and/or gladiatorial combat in the Colosseum.

*Below: Arch of Titus, raised in honour of the emperor Titus in 81 AD (map nr. 37: L 4). The triumphal processions in Imperial Rome began on **Campus Martius**, the «plain of Mars», went via Circus Flaminius, through the Arch of Titus and on along Via Sacra to Capitol, where the general or emperor placed his laurel wreath at the throne of Jupiter.*

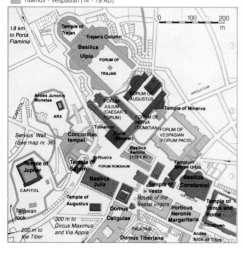

School in session in Roman Germania. Relief from Mosel valley.

Toga clad Roman seen with busts of his ancestors.

A German pays taxes to Roman tax collector.

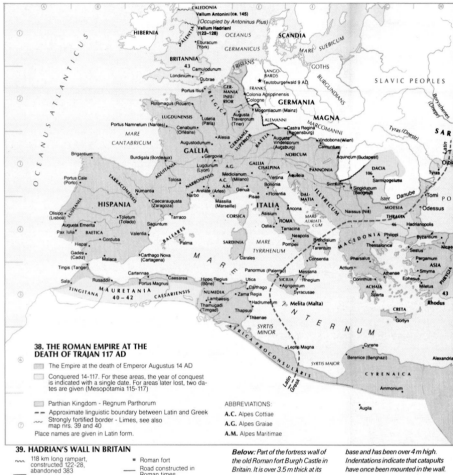

CALEDONIA
Vallum Antonini (ca. 145)
(Occupied by Antoninus Pius)
Vallum Hadriani (122-126)
HIBERNIA
•Eburacum (York)
OCEANUS GERMANICUS
SCANDIA
MARE SUEBICUM
BRITANNIA
43
Camulodunum
Londinium
•Dubrae
TRISANS
LANGO-BARDS
GOTHS
BURGUNDIANS
SLAVIC PEOPLES
Portus Itius
FRANKS
Colonia Agrippinensis Cologne
Teutoburgerwald 9 AD ★
GERMANIA MAGNA
Rotomagus (Rouen)
GERMANIA INFERIOR
Mogontiacum (Mainz)
MARCOMANNI
Borysthenes (Dnepr)
LUGDUNENSIS
Lutetia (Paris)
Augusta Treverorum (Trier)
ALEMANNI
Castra Regina (Regensburg)
Tyras (Dnestr)
SAR-
Portus Namnetum (Nantes)
Cenabum (Orléans)
•Alesia
GERMANIA SUPERIOR
Augusta Vindelicorum (Augsburg)
RAETIA
Vindobona (Wien)
Camuntum
Aquincum (Budapest)
Olbia
MARE CANTABRICUM
Augustodunum
GALLIA
•Gergovia
NORICUM
PANNONIA
DACIA
106
Tyras
Brigantium
Burdigala (Bordeaux)
Lugdunum (Lyon)
A.G.
GALLIA CISALPINA
Mediolanum (Milano)
Verona
Aquileia
Sarmizegetusa
PON-
AQUITANIA
A.C.
Bononia
Sirmium
Singidunum (Beograd)
Tomi
Portus Cale (Porto)
TARRACONENSIS
Numantia
Tolosa
NARBONENSIS
A.M.
Genua
Florentia
Pisae
DALMATIA
ILLYRICUM
Ister Danube
Odessus
HISPANIA
Caesaraugusta (Zaragoza)
Narbo
Massilia (Marseille)
ITALIA
Ancona
Naissus (Niš)
MOESIA
THRACIA
46
Hadrianopolis
Olisipo (Lisboa)
Toletum (Toledo)
Tarraco
CORSICA
Asisium
ROMA
MARE ADRIATICUM
Byzantium
Augusta Emerita
Saguntum
Ostia•
Tarracina
MACEDONIA
Philippi
Thessalonice
Nicaea
Pax Iulia
BAETICA
Valentia•
BALEARES
SARDINIA
Neapolis
Pompeii
Brundisium
Tarentum
Pharsalus
ASIA
Hispal•
•Corduba
Palma
MARE TYRRHENUM
Consentia
Actium•
Pergamum
Smyrna
Gades (Cadiz)
Carthago Nova (Cartagena)
Corinthus•
Athenae
Ephesus
Miletus
Tingis (Tanger)
Malaca
Cartennae
Caesarea
Panormus (Palermo)
Messana
•Rhegium
ACHAIA
Sparta
LYCIA
Rhodus
Sala•
Rusaddir•
Portus Magnus•
Hippo Regius (Bône)
Utica•
SICILIA
Agrigentum
Syracusae
CRETA
43
TINGITANA
MAURETANIA
CAESARIENSIS
40-42
NUMIDIA
•Carthago
•Zama Regia
•Hadrumetum
•Melita (Malta)
Gortyn
Lambaesis
Thamugadi (Timgad)
Thapsus•
MARE INTERNUM
•Thaenae
SYRTIS MINOR
•Leptis Magna
Cyrene•
•Berenice (Benghazi)
Alexandria
AFRICA PROCONSULARIS
SYRTIS MAJOR
CYRENAICA
Ammonium
•Augila

38. THE ROMAN EMPIRE AT THE DEATH OF TRAJAN 117 AD

The Empire at the death of Emperor Augustus 14 AD

Conquered 14-117. For these areas, the year of conquest is indicated with a single date. For areas later lost, two dates are given (Mesopotamia 115-117)

Parthian Kingdom - Regnum Parthorum

– – Approximate linguistic boundary between Latin and Greek

Strongly fortified border - Limes, see also map nrs. 39 and 40

Place names are given in Latin form.

ABBREVIATIONS:
A.C. Alpes Cottiae
A.G. Alpes Graiae
A.M. Alpes Maritimae

39. HADRIAN'S WALL IN BRITAIN

118 km long rampart, constructed 122-28, abandoned 383

■ Roman fort
— Road constructed in Roman times

Below: Part of the fortress wall of the old Roman fort Burgh Castle in Britain. It is over 3.5 m thick at its base and has been over 4 m high. Indentations indicate that catapults have once been mounted in the wall.

Area occupied by Agricola in 77-83, but later abandoned
CALEDONIA
Habituncum
OCEANUS GERMANICUS
Blatobulgium (Birrens)
Castra Exploratorum
Cambloglanna (Birdoswald)
Magnis (Carvoran)
Aesica (Great Chesters)
Vercovicium
Brocolitia
Cilurnum
Onnum (Halton Chesters)
Vindovala (Rudchester)
Condercum (Benwell)
Pons Aelius Newcastle
Arbeia (South Shields)
Uxellodunum (Castlesteads)
Vindolanda (Chesterholme)
Corstopitum
Segedunum (Wallsend)
Maia (Bowness)
Petriana (Stanwix)
(Old Church)
Vindomora (Ebchester)
Aballava (Burgh by Sands)
Luguvalium (Carlisle)
To the provincial capital
Eburacum (York)
(Chester-le-Street)
ITUNA AESTUARIUM (SOLWAY FIRTH)
Petrianea
Alauna (Maryport)

Roman basilica with free-standing bell tower, campanile.

Watch tower from the border with Germania. Reconstructive drawing.

Roman aqueduct – a bridge that transports water – from around 50 AD.

Right: The fortified 6.5 km long wall surrounding Constantinople, built in the time of the East Roman emperor Theodosius II, 401–450, son of emperor Arcadius.

The Roman Empire 31 BC –117 AD

The Principate

Rome is still a republic, but the head of state – *princeps* – becomes the most powerful man in the empire, since he is commander in chief of the military – *imperium* – and has the largest personal fortune.

30 BC–14 AD Octavian receives the honourary title 'Augustus'. He is caesar, i.e., emperor. Augustus expands the provinces in Spain, Gaul and along the Donau (map nr. 38) and creates peace in the Empire, *Pax Romana*, The Roman Peace.

37 Tiberius' adopted son, Caligula, as the new emperor. Four years later he is murdered by members of the Praetorian Guard.

54–68 Nero becomes an increasingly despotic, murdering several members of his own family. Rome burns in 64. The generals rebel, and Nero commits suicide.

68–69 The Year of the Four Emperors. Generals in the outlying areas of the Empire declare themselves «Caesar» and advance on Rome.

98–117 The Empire is at the height of its expansion following Trajan's conquest of Dacia (Romania), Arabia, Armenia, Assyria and Mesopotamia (map nr. 38).

Above: Two officers of the Imperial Guard. From a marble relief from the time of Hadrian, i.e., the beginning of the 1st century AD.

Britain 55 BC–128 AD

55 and 54 B.C The Roman proconsul in Gaul, Julius Caesar, crosses the Channel and attacks the Celtic tribes in the south (map nr. 33, p. 25).

43–128 AD Emperor Claudius initiates the conquest of the country. This is completed by governor Agricola in approx. 80 AD, and the island is incorporated into the empire as the imperial province of Britannia (map nr. 38). Agricola builds a breastwork between Forth and Clyde as a defense against the Scottish Picts. Under Emperor

Hadrian this line of defense is pulled back to Solvay and Tyne where a 118 kilometre long fortification, Hadrian's Wall (*Vallum Hadriani*), is constructed between 122 and 128 AD (map nr. 39). The Romans are forced to abandon the wall in 383.

40. THE ROMAN EMPIRE'S BORDER (LIMES) WITH GERMANIA CA. 100

Limes
Fort
Trade route to Scandinavia

The dragon quickly became a common motif in all Chinese art.

Chinese dual arches from Zhao period. Reconstruction.

Chinese writing consists principally of these nine brush strokes.

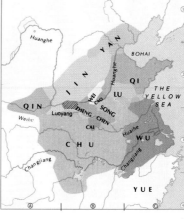

41. CHINA CA. 4000-1600 BC

— Extent of the Yangshao culture ca. 4000-3000 BC
— Extent of the Langshao culture ca. 1600 BC
▨ Approximate extent of the Shang Empire 1600-1000 BC
▲ Archaeological finds from Shang period
Names of modern provinces

China ca. 1600–206

Approx. 1600–1027 BC China's historical period begins with the Shang Dynasty, a slave-society at the height of China's bronze age (map nr. 41). A written language is developed for use by the oracular priests. A very highly advanced technique is used in the production of bronze vessels. Ancestor worship demands large numbers of human sacrificial victims.

1027 The Shang Empire is conquered by the Chou, a people from the Wei river valley in Shan Xi.

1027–771 Under the Chou Dynasty China's emperors take the title *Tian-Zi*, Son of Heaven. Slave society is succeeded by feudalism. The kings confer fiefdoms on clan

42. CHINA CA. 500 BC. THE EASTERN CHOU.

▨ Areas retained by the Chou princes around Luoyang
Not all small states are marked

and tribal leaders in return for military support. The first canals and irrigation systems are built, and new agricultural implements of iron are used. The copper coin becomes the standard of exchange. On the whole, there is a high level of cultural activity.

771–221 The eastern Chou. The period sees a bitter and bloody power struggle between rival princes. This continues until the kingdom is once again united under the first emperor of Ch'in. Most of the petiod covers the age of Spring and Autumn, *chunqiu*, 722–481 BC (map nr. 42) and the 'Epoch of the Warring States', *chan kuo*, 481–221 BC. The periods are named after two chronicles describing their histories.

221 After the Ch'in ruler in the years 230–221 BC has subjugated the other states, he takes the title Ch'in Xi Huangxi – the First Emperor of Ch'in.

221–206 Under the Ch'in Dynasty, the systems of writing, and weights and measures are standard-

Left: Cheng Xi Huangzi, the first emperor of Ch'in (map nr. 43), ruled China's first empire from 221–210 BC. The extent of his power is demonstrated by the 'imperial guard', life-size terra cotta figures, 7 000 strong that were placed in his mausoleum. This army was discovered during the digging of a well in 1974 and is one of the most sensational archaeological finds of this century. The terra cotta army had originally stood under a wooden construction, covered with earth, but after this collapsed around 207 BC, the entire imperial guard lay hidden for over 2 000 years. The figures were originally painted red, green and brown. The soldiers are all facing east, toward areas conquered by the first emperor. The warriors were fully armed, but everything made of wood, such as spear shafts and bows, have all rotted away. .

Chinese symbols for the 5 elements: wood, fire, earth, metal, water.

Sacred bronze vessel, 36.5 cm high, from the Shang period.

43. FIRST CHINESE EMPIRE

▨ Extent under the Ch'in 221 BC

▨ Expansion during Han Dynasty (see also map nr. 44)

∼∼∼ The Great Wall, begun in the 4th century BC

ized. The Great Wall (see illust. below) is built. All books in the empire are to be burned. This calamity befalls the writings of Confucius as well, the individual who played the greatest role in the development of Chinese culture. Yet it seems the book burnings were not as effective as emperor and his officials had hoped. In 210 the First Emperor dies during a tour of inspection.

206 Four years after Cheng's death, general Liu Bang, seizes power. Becomes the first emperor of the western Han Dynasty, as Gao Tzu.

Above: A bronze statuette of two wrestlers, crafted in China during the Chou Dynasty, i.e., in the last millennium BC. The statuette is now in the British Museum, London.

Left: Part of the ca. 2400 km long fortification begun in China in the 4th century BC to defend against the Huns (maps nrs. 43 and 44). The Great Wall was completed by Cheng Xi Huangxi near the end of the 3rd century BC. The world's largest construction. Much reduced in size in the 14th and 15th centuries, it is now 6–10 m high, 4–6 m thick at the top, and has some 24000 watchtowers.

Chinese warrior, dressed in trousers and short-waisted jacket.

Chinese clash in the Zhou period. Decoration on vase.

Silk production in the Tang period. Here raw silk is processed.

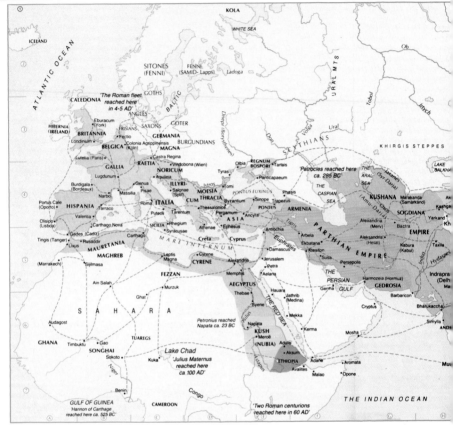

KOLA

WHITE SEA

ICELAND

Ob

ATLANTIC OCEAN

SITONES (FENNI) FENNI (SAMID- Lapps) Ladoga

GOTHS

'The Roman fleet reached here in 4-5 AD' CALEDONIA

ANGLES

URAL MTS

Tobol

Irtisch

HIBERNIA (IRELAND) BRITANNIA FRISIANS SAXONS GERMANIA GOTER

Eburacum (York)

Londinium Fectio Colonia Agrippinensis (Köln) BURGUNDIANS MAGNA

Volga Ural

KHIRGIS STEPPES

BELGICA Castra Regina

Lutetia (Paris) RAETIA Vindobona (Wien) NORICUM

GALLIA Genua Aquileia

Lugdunum ILLYRI- Istros

Dnjepr (Borysthenes)

Olbia REGNUM BOSPORI Tartais

SCYTHIANS

LAKE BALKHA

Burdigala (Bordeaux) Pisae MOESIA Tyras Tomi

Don

'Patrocles reached here ca. 285 BC' THE ARAL (Syr-Darja) Jaxartes

Narbo Massilia ITALIA CUM THRACIA Byzantium

Portus Cale (Oporto) HISPANIA Tarraco Roma Puteoli Pergamum PONTUS EUXINUS Sinope Trapezus Phasis

Valentia Tarentum ASIA Ephesus ARMENIA

THE CASPIAN SEA

THE ARAL SEA (Amu-Darja) KUSHANA Marakanda (Samarkand)

Kashga

Olisipo (Lisboa) SICILIA Rhegium Athenae Antiochia

Tigris

SOGDIANA Alexandria (Merv) Bactra Yarkand Kh

Gades (Cadiz) Carthago Nova Creta Cyprus Euphrates Arbela Ekbatana Aleksandria (Herat) EMPIRE

Tingis (Tanger) Rusaddir Carthago MARE INTERNUM Damascus Ktesion PARTHIAN EMPIRE Kabura (Kabul) Taxila

Lixus MAURETANIA Leptis Magna Cyrene Jerusalem Persepolis THE PERSIAN Indus

MAGHREB Syracusae Petra Harmozeia (Hormuz) Indrapra (Delh

(Marrakech) Sijilmasa FEZZAN Alexandria Aelana GEDROSIA Hydaspes

Ain Salah Memphis GULF Barbaricon

Ghat Murzuk Syene THE RED SEA Cryptus Bharukaccha

S A H A R A Thebae Hauara Jathrib (Medina)

Napata ca. 23 BC

Petronius reached Mekka Simylla

Audagost Napata Kerma Mosha ANDI

GHANA Timbuktu Gao KUSH Meroë

SONGHAI Sokoto Kuka 'Julius Maternus reached here ca 100 AD' (NUBIA) Aduiis Muł

Lake Chad Adane Aromata

Benin ETHIOPIA Avalites Opone

Congo Maiao

GULF OF GUINEA CAMEROON 'Two Roman centurions reached here in 60 AD' THE INDIAN OCEAN

'Hannon of Carthage reached here ca. 525 BC'

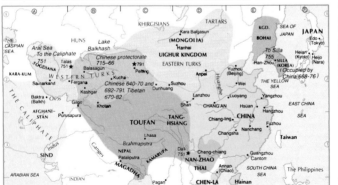

KHIRGISIANS TARTARS

KGD. SEA OF JAPAN JAPAN Edo- (Tokyo)

THE CASPIAN SEA Aral Sea HUNS Lake Balkhash Kara Balgasun (MONGOLIA) BOHAI

To the Caliphate 751 Talas Chinese protectorate 715-66 UIGHUR KINGDOM To Silla 755 Heian (Kyoto) Heijo (Nara)

KARA-KUM SOGDIANA 751★ Balasagun EASTERN TURKS SILLA (KOREA) Occupied by China 668-76)

Samarkand Fergana •791 Pehing Yuzhou (Beijing) Han-zhou THE YELLOW SEA

WESTERN TURKS Kucha Suzhou Anpei Wei

Bactra (Balkh) Oxus Kashgar Dunhuang Chinese 640-70 and 692-791 Tibetan 670-82 Lanzhou Luoyang Yangzhou EAST CHINA SEA

Gilgit Khotan Shan CHANG'AN Hsuan Hangzhou

AFGHANI-STAN Purusapura TOUFAN TANG-HSIANG Chiang-ling CHINA Nanchang Fuzhou

Brahmaputra Changsha Taiwan

SIND NEPAL Lhasa Dali 751★ Cheng-chiang Guangzhou (Canton)

Ganges Pataliputra KAMARUPA NAN-ZHAO Annan (Chiao) SOUTH CHINA SEA The Philippines

ARABIAN SEA MAGADHA THAI Hainan

INDIAN Pagan CHEN-LA

The Parthian Empire

was founded by an Iranian-Scythian nomadic people under king Arsaces in the middle of the 3rd century BC. By about 100 AD they had conquered most of present-day Iran and Afghanistan and areas of northwestern India (map nr. 44). In the middle of the 3rd century BC Parthia became part of the Sasanid Empire (map nr. 50, p. 36).

45. CHINA AROUND 750 (TANG DYNASTY)

The Empire of China under the Sui and Tang

Chinese protectorates

The Turkish kingdom (Eastern and Western Turks). Chinese vassal states.

TOUFAN Chinese vassal state

Horseman from the Han dynasty. Ceramic figure from 1st century A.D.

Model of a gatehouse from the Han dynasty, found in a grave.

Camel carrying silk along the Silk Road. Statuette from the Han dynasty.

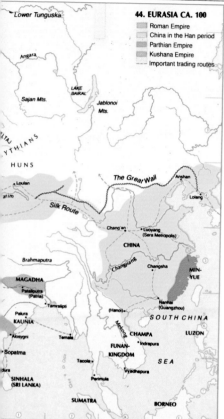

44. EURASIA CA. 100

- ▨ Roman Empire
- ▨ China in the Han period
- ▨ Parthian Empire
- ▨ Kushana Empire
- --- Important trading routes

Lower Tunguska
Angara
Sajan Mts.
LAKE BAIKAL
Jablonoi Mts.
ALTAI
SCYTHIANS
HUNS
Loulan
Tarim
Silk Route
The Great Wall
Anshan
Chang'an
Luoyang (Sera Metropolis)
Lolang
CHINA
Brahmaputra
Changjiang
Changsha
MIN-YUE
MAGADHA
Pataliputra (Patna)
Tamralipti
Nanhai (Guangzhou)
SOUTH CHINA
Pakura
KAUNIA
Alosygni
Temala
(Hanoi)
CHAMPA
LUZON
Sopatma
Tacola
FUNAN-KINGDOM
Indrapura
SEA
SINHALA (SRI LANKA)
Perimula
Vyadhapura
SUMATRA
BORNEO
Mekong

China 206 BC–907 AD

206 BC–220 AD Under the Western and Eastern Han Dynasties, Confucianism emerges victorious, and civil servants must pass examinations in Confucian literature. The caravan routes to the west are opened up, and internal trade is greatly increased. In the year 65 AD Buddhism comes to China and gradually becomes a strong influence on Chinese art. In about 100 AD paper is invented, and in 110 the first Chinese dictionary is completed.

220–289 The Time of the Three Kingdoms (Wei, Wu, Shu). The kingdom is divided. Nomadic tribes settle in the vicinities of the cultural centres and adopt Chinese culture and manners. The most important written teachings of Buddhism are translated into Chinese, again greatly influencing Chinese artists.

262 For the first time tea is drunk at court.

311 The Huns return to China's borders. Due to internal struggles within the empire, they manage to conquer all of northern China.

569–618 China is united again under the Sui Dynasty. Extensive construction is begun on canals, irrigation systems and roads. One million workers dig a canal from the river Chang-yang to the capital city of Lo-yang near the Huang-ho. Strong Chinese cultural influence is felt in Japan.

Above: Tai Zong, one of the greatest emperors in Chinese history, governed 626–49. Seen here with ladies of his court, contemp. painting of emperor during an audience.

618–907 Tang Dynasty aims to reestablish and expand the old empire (map nr. 45). Chang'an (map nr. 46) is a powerful cultural and political centre with two million inhabitants. An age of greatness for most branches of the arts. It is, among other things, the first period of blossoming in the art of calligraphy.

Below left: Ladies of the Chinese imperial court, as depicted in a mural from the Tang period. The artists of this time showed an impressive ability to capture individual character in portraits.

Below: The capital of the Sui and Tang emperors, Chang'an in N. China. The imperial palace with its harem is in the North part of the city. To the South are government offices.

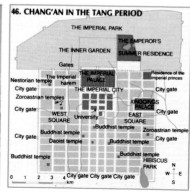

46. CHANG'AN IN THE TANG PERIOD

THE IMPERIAL PARK
THE INNER GARDEN
THE EMPEROR'S SUMMER RESIDENCE
Gates
Nestorian temple
The Imperial harem
THE IMPERIAL PALACE
Residence of the imperial princes
City gate
Zoroastrian temples
THE IMPERIAL CITY
City gate
City gate
XINGQING PALACE
City gate
WEST SQUARE
University
EAST SQUARE
Buddhist temple
Zoroastrian temple
City gate
Buddhist temple
City gate
Buddhist temple
Daoist temple
Buddhist temple
Buddhist temple
City gate
Buddhist temple
HIBISCUS PARK
N
W E
S
0 1 2 3 4
km
City gate City gate City gate

Chinese civil servant from Ming dynasty, standing before his emperor.

 All Asia feared Kublai Khan's Mongol horseman in the 13th century.

   A Japanese Samurai putting on his complicated armor.

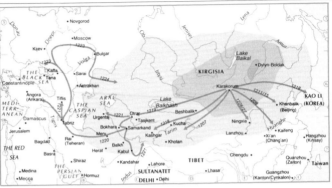

47. GENGHIS KHAN'S EXPANSION 1207-27

▨ Original power base of Temujin (Genghis Khan)

▨ Dominated by Genghis Khan (Great Khan from 1206) in 1207

▨ Mongol Empire at Genghis Khan's death, 1227

→ Mongol Campaign

See also map 55

fied period of time.
1333 The residence of the Shogunate is moved to Kyoto.
1542 The Portuguese arrive as the first Europeans in Japan.

Mongolia 1196–1227

1196–1206 The chieftain Temujin unites the Mongolian tribes, chooses Karakorum as capital and as supreme ruler takes the name Genghis Khan.
1207–1227 Western Hsia and the Jin kingdom in China are conquered. Beijing (Peking) becomes Khanbalik. Khorezm on the Caspian Sea is also taken.
1227 Genghis Khan dies. The kingdom is divided among four sons under Oghotai in Karakorum. He continues the conquest of China. The other sons push on to Russia and Eastern Europe, Iran and Mesopotamia.

China 907–1644

907–960 The Five Dynasties. Political strife and dissolution.
960–1279 Song Dynasty. China is reunited and Kai-feng becomes the capital. Gun powder is used for military purposes. Contact with countries to the west is broken off. The empire is threatened by nomadic tribes, invading from the north.
1126 N. China is conquered by the Tungu and Khitan peoples. The Song Dynasty remains intact in the south.
1251–1280 The Mongols led by Kublai Khan conquer the Song, and from 1280 he is recognized as emperor of China.
1280–1368 The Mongols now make up only a small ruling class.
1307 Foreign religions gain a foothold. The first Roman Catholic archiepiscopal see is established in Beijing (Peking).
1355 Growing dissatisfaction with Mongolian rule leads to rebellion.

1368 Mongolian rule ends.
1368–1644 The Ming Dynasty. Trade relations with Russia and countries in the west are reestablished. The Great Wall is restored (illust. p. 31) and the aqueducts repaired.
1421 Rebuilt Beijing new capital (p. 35).
1514 The first Portuguese ship arrives in China.
1557 Portuguese in Macao granted trade monopoly.
1629 The Manchus breech the Great Wall and occupy the Liaotung peninsula.
1644–1912 The Qing Dynasty.

Manchus expand their rule to Mongolia, Turkistan, Tibet and Burma. Economy is growing while the population is rising quickly.

Japan ca. 1000–1542

Ca. 1000 Height of the Fujiwara Period. Warrior-aristocracy.
1185 Minamoto Yorimoto formed a military rule with headquaters in Kamakura. The age of the Samurai commences. The emperor in Kyoto loses all real political influence.
1192 Minamoto becomes *shogun,* supreme commander, and is given dictatorial powers for an unspeci-

Above: This section from a miniature from the 1200s is probably meant to represent the Mongol lord Genghis Khan (1161–1227), who built up one of the world's most effective military forces. The mobility of the Mongol cavalrymen allowed them to suprise opponents.

Right: Torii at the Itsukushima temple, symbol of a 1000 year-old trad. According to tradition, pilgrims sail through the torii, under water at high tide. On an island southwest of Hiroshima.

Large animals guard the tombs of the Ming emperors. Here a resting elephant.

Chinese floor vase from the Ming period. 70 cm high.

Three boon companions. Detail from a 16th century Chinese painting.

China's New Capital

In 1421 Emperor Yung Lo moved China's capital from Nan-king to Beijing (Peking) to bring his administration nearer to the border with Mongolia.

In the struggles against the Mongols, much in Beijing, or Khanbalik as it was known then, was destroyed. Now a number of new and magnificent buildings were constructed, both in the Forbidden City, where only the emperor and his attendants were allowed, and in the Imperial City.

On the site of the city's sacrificial alter the Temple of Heaven was erected. **The picture at the left** shows the core of this complex. Because the temple is roofed with blue tile, it is also referr to as the «Blue Temple». In spring the emperor sacrificed and prayed for a good harvest. This remnant of China's prehistory was revived during the Ming Dynasty and practised right up until the Revolution in 1911/12. Today the temple complex is a great tourist attraction.

Following a period of restoration in the 1420s, Beijing consisted of four walled cities, one inside the other. Innermost was «The Forbidden City», then «The Imperial City», which housed the administration, then the «Tartar City» and, finally, the «Chinese City». The perimeter wall was 20 m high and 30 km long.

48. CHINA DURING THE MING (1368-1644) AND QING (1644-1912) DYNASTIES

Chinese Empire ca. 1500 (Ming)
Chinese Empire ca. 1800 (Qing)

Sasanid king Arda-shir III (628–30) on a deer hunt.

Crowns worn by Sasanid kings Ardashir I and Shapur I.

The various stances and prostrations during Muslim prayer.

Above: No one can say exactly how early Buddhism reached the Mon people of Lower Burma, but according to their own legends the brothers Tapusa and Palikat are supposed to have personally received eight hairs from the head of the Buddha. The brothers brought this gift back to their homeland near present-day Rangoon, where they built the shrine Shwe Dagon to contain the hairs. The gilded pagoda still exists. This lithograph, however, shows the shrine overgrown with tropical vegetation from the beginning of the 1800s. From a colour lithograph by J. Moore from 1825.

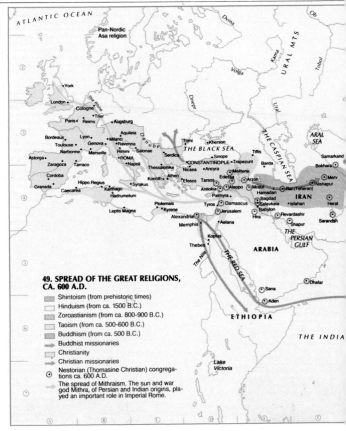

49. SPREAD OF THE GREAT RELIGIONS, CA. 600 A.D.

- Shintoism (from prehistoric times)
- Hinduism (from ca. 1500 B.C.)
- Zoroastianism (from ca. 800-900 B.C.)
- Taoism (from ca. 500-600 B.C.)
- Buddhism (from ca. 500 B.C.)
- → Buddhist missionaries
- Christianity
- → Christian missionaries
- ⊙ Nestorian (Thomasine Christian) congregations ca. 600 A.D.
- The spread of Mithraism. The sun and war god Mithra, of Persian and Indian origins, played an important role in Imperial Rome.

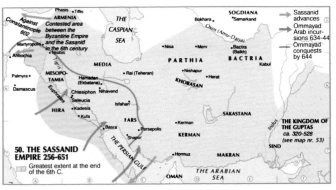

50. THE SASSANID EMPIRE 256-651

Greatest extent at the end of the 6th C.

→ Sassanid advances
→ Ommayad Arab incursions 634-44
→ Ommayad conquests by 644

THE KINGDOM OF THE GUPTAS ca. 320-528 (see map nr. 53)

Members of a caravan. Detail from an Arab miniature.

Floor-plan from 1600s of the Muslim holy of holies – the Kaaba in Mecca.

Muhammed holds his farewell seremon. Arab miniature.

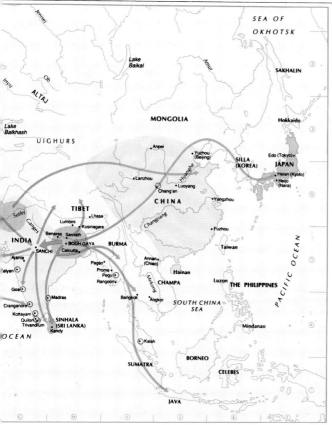

The Arabian Empire
622–732

622 The founder of the Islamic religion, the prophet Mohammed (b. 570) flees Mecca for Medina – an event that later came to mark the beginning of the Islamic calendar.

630 Mohammed returns from Medina to Mecca following many years of strife between the two cities. He dies two years later.

632–634 The first caliph, Abu Bekr, unites Arabia and puts down all attempts at division.

634–644 The second caliph, Omar I, conquers large portions of the East Roman (Byzantine) and the Sasanid empires. In 635 Syria is taken, in 637 Mesopotamia, in 640 Palestine and in 642 Egypt (map nr. 51).

650 The third caliph, Othman, conquers the remainder of the kingdom of the Sasanid (Persia).

661 When the fourth caliph, Ali, is murdered, the unity of Islam is lost, and the Shi'ites refuse to accept subsequent caliphs as legal rulers. The Omayyad clan comes to power. Damascus becomes the new capital.

705–715 The caliphate of the Omayyad reaches its zenith during the reign of Walid I. Following the conquest of North Africa, commander Tarik Ibn Ziyad crosses the Strait of Gibraltar and begins the conquest of the Iberian Peninsula. In the east, the western part of the Indus valley is conquered in 710-713. The Arabs push on through Afghanistan and cross the Oxus.

717–718 The Arabs lay siege to Constantinople, but are forced to withdraw.

732 Arabs defeated at Poiters by the Frankish ruler of Austrasia, Charles Martel.

The Sasanid Empire

the beginning of the 3rd century the powerful Parthian Empire (map nr. 44, p. 32) was on the verge dissolution. A clan that had ginally come from Fars in south-st Iran – the Sasanid - managed a very short time to create an-er Persian great-power (map nr.

e first leader of note of the sanid dynasty was king Ardashir who reigned from 224 to 241. A cture of him is carved out in a cliff ief in Firuzabad (see illust. left). e Sasanid epoch in Iran 6-651) was a rich one for the ion, both politically and cultur-y, and Zoroastrianism became e official state religion.

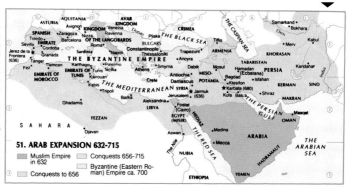

51. ARAB EXPANSION 632-715

- Muslim Empire in 632
- Conquests to 656
- Conquests 656-715
- Byzantine (Eastern Roman) Empire ca. 700

A fearsome Indian war elephant from the 3rd century AD.

The Stupa in Sanchi (see also the illustration below).

The god Shiva, dancing. S. Indian representation, cast in bronze.

52. THE MAURYA KINGDOM IN INDIA

▨ Greatest extent ca. 250 B.C.

— Alexander the Great's campaign 329-325 B.C. (see map 26)

Below: In 1817 in Ajanta in western India (map nr. 53), British soldiers found several monasteries and temples that had been carved into the side of a mountain. Originally there had been only one cave, dedicated to the serpent god Naga, but several more were added, until today there are twenty-eight in all. Most are decorated with murals. A section of one of them is reproduced here – a religious, erotic art form, which influenced Asian art for centuries.

India 329 BC–646 AD

329–325 BC The kings in the greater empire are engaged in internal struggle, and when Alexander the Great pushes into India from Afghanistan in 326, he meets an unified resistance. Some actually become his allies, while others are vanquished. In 325 Alexander's troops refuse to follow him any farther east. He is forced to return home (see also p. 21). At Alexander's death in 323 the Hellenic colonies have founded revert to Seleukus I of Syria.

Ca. 320 The Maurya Dynasty, led by Chandragupta Maurya, comes to power and quickly regains control of the areas Alexander had conquered.

Ca. 305 Seleukus I campaigns in India in order to protect his possessions, but is defeated by Chandragupta.

Ca. 270 The Maurya Dynasty's greatest ruler, Ashoka, comes to power under bloody circumstances. Legend tells that he had first to dispose of ninety-nine brothers!!

Ca. 261 Ashoka attacks and conquers the state of Kalinga. The knowledge that the battles have cost the lives of 150 000 men puts the king in a state of despair. He becomes a believing Buddhist and proponent of non-violence, commanding his subjects to show moderation, friendliness, tolerance and piousness. He sends Buddhist missionaries to Greece, Syria, Egypt, Sri Lanka and Indo-China.

Ca. 235 Ashoka dies. The kingdom gradually splits up into several smaller states.

Ca. 320–528 AD The Gupta Dynasty. Northern India is united again under Chandragupta I in 320. The dynasty holds power for two hundred years, with the empire reaching the limits of its expansion under Chandragupta II (map nr. 53). The period is a golden age in art and literature.

606–647 After the country has been severely weakened during an attack by the White Huns in the middle of the 400s, king Harsha manages for a short time to hold together a kingdom in the north (map nr. 53). He makes Kanauj his capital and assembles scholars and poets at his court.

53. INDIA UNDER GUPTA DYNASTY AND HARSHA

▢ Gupta Empire 385-414

➡ Attack by White Huns, mid-5th C.

▨ Harsha's kingdom 606-46

Woman from the Khmer kingdom. Detail of relief from the 12th century.

Representation of the Buddha from the Khmer kingdom. From the 12th century.

Detail depicting mythological beast – a makara – from central Vietnam.

Left: Shortly after the Buddha's death in 485 BC, he was revered as a god. His corpse was burned, his ashes kept as holy relics. Stupas were built over these relics. This picture shows a stupa in Sanchi in India, erected in the last century BC. It is 36 m in diameter and 16.5 m high. The stupas later became more like East-Asiatic temples. The tops of the mounds were built out so that the stupas became pagoda-shaped (illus. p. 36).

Right: At the zenith of the Khmer Empire this beautiful head of the Buddha was created. It was carved in sandstone in the 1100s and is now in the Musée Guimet, Paris.

Cambodia (the Khmer Kingdom) 600–1200

Approx. 600 The Khmer people from the highlands in the north advance into Funan and take power there.

The 700s The kingdom is for a time a dependency of Shailendra – a larger kingdom, centred in Java and Sumatra.

802–850 Jayavarman II liberates the Khmer Kingdom from Javan domination and makes Angkor the religious centre of the kingdom.

The 1100s The Khmer Kingdom's period of greatness. Under Suryavarman II (1113–1150) attains its greatest expansion (map nr. 54), and it is in his day that work on the mighty Angkor Vat is begun (see illust.). Near the end of the century Jayavarman VII, who governs the kingdom 1181–1218, erects Angkor Thom, a huge fortress, built around a temple.

Approx. 1200 Following continuous attacks from the Thai people and internal strife, the kingdom is in a state of disarray.

Right: Angkor Vat is the largest of the many temples in the Khmer Kingdom's capital of Angkor. It is thought that construction on the temple complex was begun by king Suryavarman II and completed later. It covers an area that is roughly rectangular, with a front side 185 m long. The main tower in the centre is 65 m high, and surrounded by four smaller towers, one in each corner. Angkor Vat is considered one of the most important surviving architectural structures from the Khmer Age in Cambodia.

54. INDOCHINA AND THE KHMER KINGDOM DURING THE ANGKOR DYNASTY

CHINA

VIETNAM (ANNAM)
Independent from 939
LANNA
Hanoi
PAGAN
TOUNGOO
Chiang Mai
Prome
Vien Chang
Hainan
Pegu
Sukhothai
Thaton
Irrawaddy
Menam
Mekong
Indrapura
KHMER KINGDOM
(KAMPUCHEA)
CHAMPA
Ayutthaya
ANGKOR
Vijaya
Andaman Islands
FUNAN
Phan Rang
Phnom Penh
SOUTH CHINA SEA

Nicobar Islands

Samudra

Perlak
PAHANG

SUMATRA

☐ Kingdom ca. 800

▨ Greatest extent under Suryavarman II (1113–50)

Mongol horseman didn't just wage war – they played polo too.

One of the Mongol era's last emperors on the Chinese throne – Wen Zong.

A Mongol and his indispensable steed in a strong wind.

55. EURASIA CA. 1300. THE SPREAD OF CHRISTIANITY AND ISLAM

- Mongol dominion
- Christian dominion
- Islamic dominion
- Important trade routes

Mongol Domination
1227–1405

1227 The Mongol Empire is divided, becoming four smaller kingdoms under the supreme command of the Great Khan Oghoai (see p. 34): The Golden Horde, north of the Black and Caspian Seas, the kingdom of the Il-khan in Iran and Afghanistan, the Chagatai-khanate and the kingdom of the Great Khan. Most Russian fiefdoms too come under Mongol control, the so-called «yoke of the Tartars».

The Kingdom of the Great Khan

1229–1241 Genghis Khan's son, the Great Khan Oghotai, insures the dominion of Northern China and makes Karakorum the capital.
1267–1279 Kublai-Khan, Oghotia's nephew, conquers the Song Empire in southern China, moves the capital to Dadu (Beijing) and calls it Khanbalik (city of the Khan). He becomes emperor of China in 1280.
1280–1368 Kublai-Khan attacks Toungoo (Burma) and conquers Pagan (1287), but the Mongols are

eventually driven back.

The Golden Horde 1237–1241

Genghis Khan's grandson, Batu, regent in the khanate of the Golden Horde, attacks Europe. Kiev is conquered, Moscow is sacked and burned. The Mongols are victorious in sixtyfive battles, and in Europe it is feared that it is only a matter of time until the plundering hordes from the East have conquered the entire continent. In 1241 it is Vienna's turn, but right in the middle of the siege, Batu receives word of Oghotai's death. He breaks

off at once so that he can return to take part in the succession question in Karakorum.

The Khanate of Chagatai

1227 Genghis Khan's second son, Chagatai, becomes khan and makes his own khanate the political centre of the Mongol kingdoms.

The Empire of the Ilkhans
Approx. 1237–1258

The Mongols get a foothold in Iran, and in 1258 Genghis Khan's grandson, Hulagu, sacks Baghdad. The caliph is killed. For almost two

Timur-lenk on the throne. Copied from Persian miniature from the 1600s.

Mosque lamp of glass with quotations from the Koran. From the 1300s.

A mosque in Tunisia, built in the 8th century.

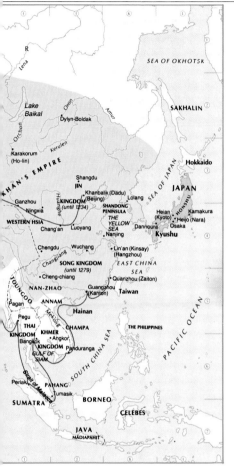

56. THE SPREAD OF CHRISTIANITY IN EUROPE

→ Missionaries
▦ Greek Orthodox area
▦ Roman Catholic area
▦ Anglo-Saxon area
● Archbishopric with year of establishment
▦ Islamic area

Below: Islam extended itself in the 1200s (map nr. 55), and Arab traders were common in both the East and the West. Here two travellers on camels are received at the city gate. From an Arab miniature from 1237.

hundred years the Ilkhans hold power in Iran, Iraq, Afghanistan and parts of Asia Minor.

The Dissolution of the Mongol Empire 1369-1405 Timur-lenk attempts to establish a new empire. From his base in Samarkand, his troops invade the kingdoms of the Ilkhans, which they conquer. He then pushes into the khanate of the Golden Horde and into northern India, but the newly founded great kingdom unravels quickly when Timur-lenk dies in 1405.

Spread of the World's Major Religions

Map nr. 56 Boundaries of the Catholic faith in early medieval times. Missionizing from Christendom's two great centres, Rome and Constantinople, from 1054 divided in a Roman Catholic and a Greek Orthodox Church.
Map nr. 55 shows the areas claimed by the Christian church by about 1300, and how far Islam had spread at the same point in time.
Today only Christianity has more adherents than Islam.

Byzantine fishermen. From a 6th century mosaic.

The Byzantine Imperial Palace in Constantinople in the 6th century.

Byzantine warrior from the beginning of the 7th century.

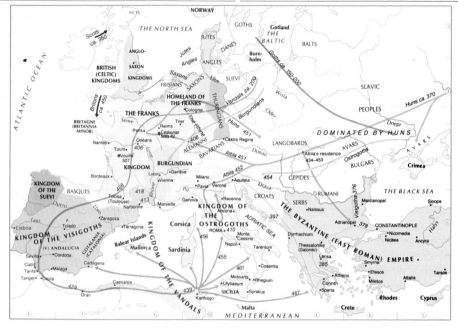

The Byzantine (East Roman) empire 395–626

58. MEDITERRANEAN LANDS CA. 600

— The Byzantine Empire at the death of Justinian the Great in 565

The Byzantine Empire ca. 600

Kingdom of the Lombards from 568

The Suevi Kingdom, lost to Visigoths in 585

395 Emperor Theodosius dies, and as Rome's size makes it difficult to administrate from one central point, his sons each become emperor of one half of the empire: Eight-year-old Honorius in the West Roman Empire, his seven-year older, and equally ungifted, brother, Arcadius, in the East Roman

(Byzantine) Empire. Constantinople becomes the capital in the east, with Roman law and administration, but Greek language and culture.

527–565 Under Emperor Justinian I, North Africa, Italy and southern Spain are conquered (map nr. 58).

568 The Langobards take most of Italy, and Byzantine outposts in Ravenna.

626 Constantinople is attacked by Sasanidians and Slavs.

Below: The wife of Emperor Justinian I, Theodora surrounded by clergymen and ladies in waiting. From a mosaic in the church of San Vitale in Ravenna. The costume of the Byzantine church and court is embroidered with gold and decorated with jewels and pearls.

German knight from the 700s with short sword and long lance.

Frankish warrior from the 600s reaching for his dagger.

Gilded helmet from the middle of the 7th century, found in East Anglia.

57. GERMANIC MIGRATIONS AND THE GERMAN KINGDOMS 526

→ Vandals   → Visigoths
→ Goths   → Burgundians
→ Ostrogoths → Huns
⇢ Jutes, Angles and Saxons

Right: From the imperial palace in Constantinople there was a direct passageway to the imperial loge in the Hippodrome, where not only horse racing, but also mystery plays, mock hunts and acrobatics could be seen. This section of a palace mosaic from the latter half of the 6th century shows two athletes attacking a wild animal.

The Germanic Migrations

The Goths are the first to migrate, from Scandinavia, around 150 AD, reaching the coast of the Black Sea around the end of the century. At about the same time, the Burgundians and Vandals move westward. The former establish the Burgundian Kingdom in southeastern France in about 400. The Vandals found a kingdom in Spain, (V)Andalusia, at about the same time, but continue on to North Africa. The Goths in the vicinity of the Black Sea, who by the end of the 3rd century had separated into Visigoths (Western Goths) and Ostrogoths (Eastern Goths), also migrate toward the rich areas in the west. In Spain and the southernmost part of France the Visigoths establish their kingdom near the end of the 5th century. The Ostrogoths found a kingdom in Italy (493–553).

Kingdom of the Franks

Ca. 350 Franks migrate into Gaul.
486 King Clovis defeats the Roman governor Syagrius at Soisson, putting an end to Roman rule.
507 Clovis subjugate Visigoth territory north of the Pyrenees.
508 Capital moves from Soisson to Paris.
567 Kingdom divided into Autrasia, Neustria and Burgundia.
714–717 Charles Martel, grandson of Pipin completes the unification of the kingdom, begun by Pipin.
732 At Poitiers, Charles Martel puts a stop to the Moorish invasion over the Pyrenees.

59. WESTERN EUROPE CA. 750

▭ Kingdom of the Franks in 741
▭ Territories allied with the Franks
▭ Kingdom of the Lombards in 744
▭ Byzantine possessions
▭ The Muslim Empire

Charlemagne (an il-literate) made only two marks in centre of his signature.

A catapult bombards the fortress with stones.

Charlemagne's pal-ace, Ingelheim near Mainz. Reconstruc-tive drawing.

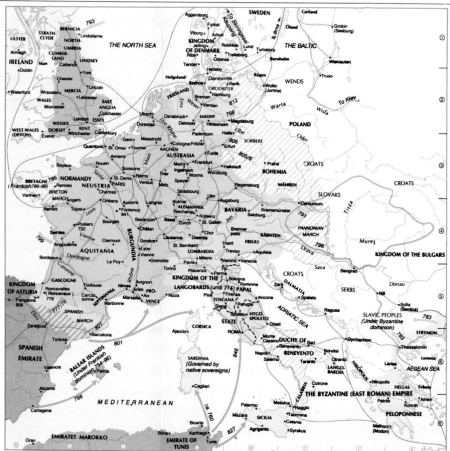

60. CHARLEMAGNE'S EMPIRE 814

- The empire in 768
- At Charlemagne's death in 814
- Arab states in 814
- The Byzantine Empire
- Kingdom of the Bulgars 802-14
- Kingdom of the Avars to 796
- Carolingian advances
- Arab incursions
- Byzantine advances
- Viking raids (see also map 63)
- Important trade routes

The Kingdom of the Franks (France) 768–843

768 Pipin the Small dies. His son, Charles (Charlemagne), succeeds him.

773–774 Charles defeats the Langobards in Italy and incorporates their kingdom. A short time later, Saxony and Friesland are also conquered (map nr. 60).

777–778 Prominent Arabs in Spain ask Charles for help, and the Franks advance all the way to the Ebro, but are forced to return.

788 Bavaria is incorporated.

800 On Christmas Day, Charles the Great is crowned Holy Roman Emperor by Pope Leo III.

814 Charles dies in January in his favourite city of Aachen and is buried in the cathedral there. Charlemagne's eldest son has died, and a younger son, Louis the Pious, assumes the office.

817 Louis the Pious' eldest son, Lothar I, becomes co-emperor.

830 Following a rebellion, Louis the Pious cedes the rank of emperor to Lothar. In the years that follow there is serious feuding between Lothar, his brother Louis the

German, and his half brother Charles the Bald.

843 Following the Treaty of Verdun, the Frankish kingdom divided in three (map nr. 62). The West Frankish Kingdom is govern-ed by Charles the Bald, the East Frankish Kingdom by Louis the German, and the Middle Kingdom by Emperor Lothar I. This division would give rise to the three nation-states of France, Germany and Italy.

A monk at his desk. From an ivory relief.

Cavalryman from the 900s, with saddle, but no stirrups.

11th century war ship with high stem in the Mediterranean.

The Byzantine (East Roman) Empire 673–1054

673–677 The Arabs lay siege to Constantinople, but are put to flight by «Greek fire» – a kind of early flame-thrower, consisting of sulphur, pitch, naphtha, lime and saltpetre.

726 Emperor Leo III bans «idolatry» (*ikonoduli*) and decrees that the «holy» images be removed from the churches. The great iconoclastic dispute rages for more than a hundred years.

751 Ravenna is conquered by the Langobards.

860 Constantinople is attacked by vikings (map nr. 63, p. 46).

907–976 Several Russian attacks on the Balkans and Asia Minor. Emperor John I Zimisces expels them in 971, retakes Syria and Palestine from the Arabs and annexes the eastern parts of Bulgaria (see next column).

976 Emperor Basil II, nicknamed «Slayer of the Bulgarians», also leads successful campaigns against the Arabs and the Bulgarians. The empire reaches the height of its power.

1054 The final break with the western church (The Great Schism).

Bulgaria 893–1018

893–927 Emperor Simon rules over most of the Balkan peninsula, with the exception of Greece, Croatia, Dalmatia and the district around Constantinople.

917–970 Under Peter I much territory is lost. The western portions of the empire become an independent Serbian kingdom in 963.

971 The eastern territories are conquered by Byzantium. The capital is moved to Ochrida (Ohrid) in Macedonia, and in **1018** the remainder of the country is incorporated in the Byzantine Empire.

61. SOUTHEASTERN EUROPE CA. 1050

- Byzantine Empire at death of Constantine IX in 1054
- Greek Orthodox see (archbishopric red)
- Roman Catholic see (archbishopric blue)
- STRYMON Byzantine thema (province)
- Arab possessions

Below: In addition to contemp. history, Charlemagne also encouraged scribes to copy older manuscripts. Here the emperor is receiving a copy of the Ten Commandments from Moses. From a miniature (ca. 850).

62. DIVISION OF CHARLEMAGNE'S EMPIRE 843

- West Frankish Kingdom
- East Frankish Kingdom
- The Middle Kingdom

Stem of a Norwegian viking ship from Oseberg. Built around 800.

Ship builders in Normandy. Detail from Bayeux tapestry.

Anglo-Saxon axe-head and sword point from viking times.

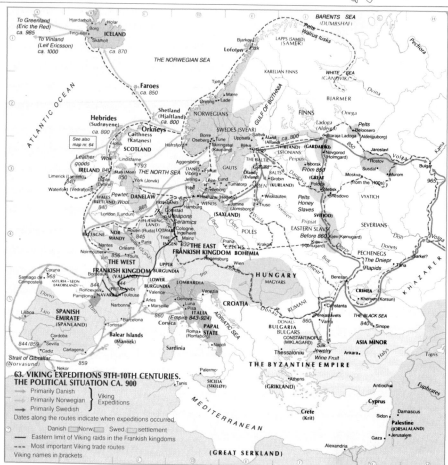

63. VIKING EXPEDITIONS 9TH-10TH CENTURIES. THE POLITICAL SITUATION CA. 900

⟶ Primarily Danish ⎫
⟶ Primarily Norwegian ⎬ Viking Expeditions
⟶ Primarily Swedish ⎭
Dates along the routes indicate when expeditions occurred.

Danish☐ Norw.☐ Swed.☐ settlement
— Eastern limit of Viking raids in the Frankish kingdoms
- - - Most important Viking trade routes
Viking names in brackets

The Vikings

We are not certain of the exact origin of the word 'viking', but we do know that from about **800 AD** they ravaged with fire and sword along the coast of Europe in their newly developed seagoing vessels. The first expedition we know of went from western Norway to Wessex in **789**. Four years later the monastery at Lindisfarne in Northumbria was attacked. All the inhabitants were killed, the cattle butchered and the cloister looted for everything of value before it was burned to the ground. These pillaging raids, in which Swedes and Danes also took part, evolved later into trading voyages, the conquest of outposts and the founding of settlements.

Around **900** there were Danish settlements in eastern England (the Danelaw), Normandy, Friesland and northern Germany. In Greenland, Iceland, the Faroe Islands, the Hebrides and the Orkneys, there were Norwegian colonies. So too in parts of Scotland and Ireland, on the Isle of Man and in Wales (Bretland). The Swedes established a number of colonies in Russia (map nr. 63).

England and Normandy 829–911

829 King Egbert of Wessex (802–839) subjugates the other minor kingdoms and becomes the first king of England. The viking attacks become more numerous and the viking fleet larger. From the middle of the 9th century large areas of England and Normandy are intermittently occupied by vikings (map nr. 64). The vikings begin taking land in Northumbria, and establish themselves as settlers in the Danelaw. **871** Alfred the Great of Wessex becomes king.

He manages to unite the other tribes in a common struggle against the vikings. In **878** he defeats them at the Battle of Edington. The ensuing peace leaves only the Danelaw to the Scandinavians, while the Kingdom of Wessex now becomes the leader of a free England. **911** Charles III of France is forced to cede a large portion of Normandy as an independent duchy to the viking chieftain Rollo.

Right: Danish vikings on their way to England. From a miniature painting in an English manuscript from the 9th century.

The cathedral in Mainz, in the beginning of the 11th century.

Man with scythe. Drawing from beginning of the 10th century.

Otto the Great of Germany accepts humble tribute from the vanquished.

The German Empire under Otto the Great 936–973

936 Otto I of Saxony, nicknamed the Great, is made king.

939 Otto puts down a rebellion among his dukes and gives the duchies to members of his own clan.

951 Having defeated the Langobard king Berengar II, and after marrying the widow of the Frankish king Lothar, Otto is crowned king of the Langobards.

955 The Magyars are defeated in the Battle of the Lechfield.

962 Otto I is crowned Holy Roman Emperor by Pope John XII in Rome. Italy is united with the German Empire.

Right: William the Conqueror's Norman vikings on horseback, attacking the newly installed Harald Godvinsson's bodyguard during the Battle of Hastings on October 14, 1066. King Harald was killed, and William became king of England. From the Bayeux Tapestry, which according to tradition was woven by William's queen, Mathilde and her ladies in waiting in the 1080s.

64. 9TH CENTURY ENGLAND

Danelaw 886
Territories occupied by Vikings

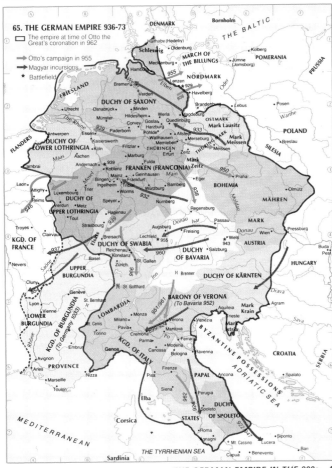

65. THE GERMAN EMPIRE 936–73

The empire at time of Otto the Great's coronation in 962

Otto's campaign in 955

Magyar incursions

★ Battlefield

«Centre of the world» – Church of the Holy Sepulchre. Christ at centre.

Even monks and nuns could land in the pillory.

A crusader takes a Muslim prisoner. Miniature from the 1200s.

66. EUROPE CA. 1100. THE FIRST THREE CRUSADES 1096-1192

Arab, mainly muslim, states

Christian lands:
- Roman Catholic
- Greek Orthodox

First Crusade 1096-99
Godfrey of Bouillon, Raymond of Toulouse, Robert of Normandy, Bohemund of Taranto

Second Crusade 1147-49 Conrad III (Holy Roman Emperor) Louis VII of France

Third Crusade 1189-92 Richard the Lionheart, Phillip II Augustus of France, Frederick Barbarossa (Holy Roman Emperor)

Left: Pope Urban II proclaims the First Crusade against the Seljuks – who had retaken the Holy Land. The conciliium decided enthusiastically that on August 15, 1096 – when the year's crops had been harvested – the crusaders would leave their respective homelands and be reunited in Constantinople.
4000–5000 knights and approx. 30000 foot soldiers set out. In the spring of 1097 those who hadn't died of hunger or sickness, or been killed in battles with the Hungarians and Bulgarians on the way, were evacuated over the Bosporus.

The Crusaders

The First Crusade (1096–1099)
There were four main points of departure. Raymond of Toulouse, who due to his wealth and skill as a commander demanded to lead the entire crusade, began in Toulouse. The pious and unselfish Duke of Lower Lothringen, Gotfried of Bouillon, set out from Stenay in northern France.
Robert of Normandy led the adventurous Normans, who made up such a large share of the crusaders that the procession nearly resembled a Christian viking expedition.

He started in Lyon.
During the siege of a city in southern Italy, the stocky Norman, Bohemund of Taranto caught sight of Robert's column as it marched past. He saw at once that this offered his adventurous spirit a much larger field of play. Breaking off the siege, Bohemund joined one of the four great columns of army of crusaders from Brindisi.
A half century later Abbot Bernhard of Clairvaux called for a new crusade, in order to come to the aid of the Christians in Syria. **The Second Crusade** (1147–1149) was led by the deeply religious

Crusader Richard the Lionheart (I) doing battle with a dangerous foe.

A soldier cocks his crossbow, an effective and accurate weapon.

An English farmer on his way to the landlord's mill to mill his grain.

Smolensk

SELJUK
EMPIRE (Kgd. 1196-1375)
Heraclaea
LITTLE ARMENIA
BARONY
Marash
Edessa
OF EDESSA
(1098-1144)
Tarsos Adana
Seleucia Antiochia
Port St. Simeon
Aleppo
CYPRUS
(Byzantine. Indp.
Kgd. 1192-1489)
PRINCIPALITY OF ANTIOCHIA
(1098-1268)
Famagusta Tortosa
(1102-1291)
Hama
BARONY OF TRIPOLI
(1102-1291)
Homs
Tripoli
(1109-1289)
Krak
Beirut
(1110-1291)
Baalbek
Sidon
(1110-1291)
Damascus
Tyros
(1124-1291)
Akka (Acre)
(1104-1187 og
1191-1291)
Hattin 1187
Tiberias
Sea of Galilee
Caesarea
(1101-1265)
Nasaret
Jaffa
Amman
Askalon
(1153-1247)
JERUSALEM
1026-1187 and
1229-1244)
Gaza Hebron
CAIRO CALIPHATE
THE DEAD SEA

Kiev

EMIRATE OF DAMASCUS

67. THE CRUSADER STATES CA. 1140
Kgd. of Jerusalem 1099-1187
Kgd. of Jerusalem after Peace of Jaffa 1229
Dates in brackets give years of crusader rule

Salei
Dnestr

THE BLACK SEA
Adrianopel Gangra
CONSTANTINOPEL (BYZANTIUM)
Nikomedeia SELJUK
Nikaia Angora (Ankara) EMPIRE
Dorylaion 1097 Caesarea
Pergamon Filomelion Marash
Smyrna Sardes Ikonion Edessa
Laodicia Tyana Tarsos
Ephesos 1148 Attaleia Seleucia Antiochia
Louis VII 1148
Rhodos Richard's fleet 1191 Cyprus Tripoli
Candia Limasol (Nemesos) Damascus
Conrad III 1148 Tyros
Philip II August 1191 Akka (Acre)
JERUSALEM
(Besieged for the first time by crusaders in the summer of 1099)
Alexandria Damietta

Glasgow
Copenhagen
Rostock Greifswald
Oxford Cambridge Frankfurt
Louvain Wittenberg
Caen Cologne Leipzig
Mainz Erfurt Praha Krakow
Paris Trier Würzburg
Heidelberg Ingolstadt
Nantes Orléans Freiburg Wien Pressburg
Poitiers Angers Bourges Basel Besançon Ofen
Dôle
Bordeaux Grenoble Torino Treviso
Cahors Padova Fünfkirchen
Palencia Toulouse Montpellier Aix Bologna
Valladolid Huesca Perpignan Firenze Perugia
Salamanca Zaragoza Lerida Siena
Avila Alcala de Barcelona Roma
Henares
Lisboa Napoli Salerno
Valencia Palma
Sevilla Catania

Louis VII of France, and Conrad III of Germany, and ended with a failed attempt to take Damascus.

The Third Crusade (1189–1192) was led by Richard the Lion-heart of England, the French king Philip II August, and the Holy Roman Emperor, Frederick I (Barbarossa). Many English and French soldiers were this time transported by sea. Although the crusade was led by the three most powerful men in the West, it ended in nothing more than small bands of unarmed Christians making pilgrimages to the Holy Sepulchre.

Right: Students following a lecture at the university in Bologna. From a miniature from the beginning of the 1400s. This university was founded as early as 1088, and during the 1100s was renowned as a school of law. In the Middle Ages the University of Bologna had as many as 2000 students from every corner of Europe. It is also from this period it has its nickname, la dotta, the learned one.

The map above right shows the founding dates of European universities in the Middle Ages.

68. UNIVERSITIES IN THE MIDDLE AGES
Founded before 1250
1251-1350
1351-1450
1451-1506

Top: During the First Crusade, 1200–1300 cavalrymen and over 10000 foot soldiers reached Jerusalem on June 7, 1099, and on June 15 the city fell. Christendom had control of the city once again, and could guard the Holy Sepulchre. From a Venetian miniature from the 1200s.

Scandinavian fisherman hauls in his catch in the Baltic in the 1200s.

Man drives his wife home from the fields, tempted by her liquor.

The baker and his wife shove bread into 2 m deep oven.

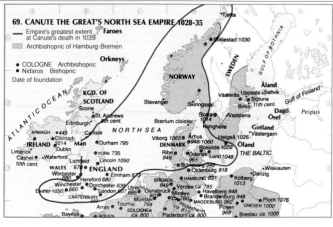

69. CANUTE THE GREAT'S NORTH SEA EMPIRE 1028-35

Empire's greatest extent at Canute's death in 1035
Archbishopric of Hamburg-Bremen

● COLOGNE Archbishopric
● Nidaros Bishopric
Date of foundation

Above: A French soldier from the 11th or 12th century, armed with sword and shield. A chess piece in ebony, now in the Bibliotheque Nationale, Paris.

The North Sea Empire 1028–1035

In 1002 king Ethelred II (the Unready) of England decreed that all Danes in the kingdom were to be killed. Sven Forkbeard, and future Norwegian King Olav Haraldsson, king of Denmark from 983, and Norway from 1000, took revenge by setting out year after year to plunder England. London was attacked in 1013, Ethelred fled the country, and Sven claimed the throne.

Sven's son, Canute the Great, became king in 1016, and conquered the whole of England. He dissolved the viking army and gave rich gifts to the Anglo-Saxon church.

In 1018 Canute succeeded his brother in Denmark, and ten years later was himself proclaimed Norway's king. His North Sea empire had become a European great-power (map nr. 69).

Frederick II von Hohenstaufen

Frederick, son of Henry VI, grew up in Sicily and spent the greater part of his life there. His foremost political ambition was to create a strong state in Italy. During the decade of the 1220s he reestablished the strong royal power of the Normans in southern Italy and on the island of Sicily, which had both seen periods with short-lived, weak governments. His attempts to accomplish something similar in northern and central Italy led to a long struggle with the pope and the Italian cities. He was excommunicated, and in Germany faced open rebellion. Frederick II had strong intellectual interests, and his court in Palermo was the cultural centre of southern Europe during the time of his reign, and a meeting place for intellectuals from far and wide.

Below: Frederick II (1194–1250), king of Sicily 1197, German king 1212, Holy Roman Emperor 1220. From a miniature in his book on falconry, now in the Vatican Museum.

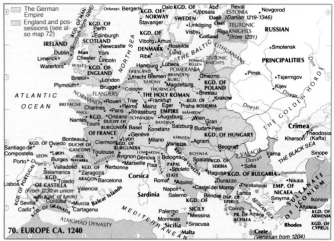

70. EUROPE CA. 1240

Palisades of wood and defence towers in old Novgorod.

Hagia Sofia in Kiev, built ca. 1018–37. Reconstructive drawing.

codifies the laws. The duke is supreme judge and commander in chief, but must consider the wishes of the Bojars (noblemen) and merchants. He goes to war with the Poles, dealing the Patzinaks a decisive defeat. The kingdom has its greatest period of expansion (map nr. 71).

After 1054 At the death of Yaroslav, the kingdom is divided among his sons. Constant warring between princes gradually reduces the kingdom to a loose confederation of principalities.

Above: A Russian village as it probably looked in the 9th and 10th century. Drawing by P.N. Tretjakov.

Russia 862–1054

862 According to tradition, the **væring**, i.e., viking, Rurik and his brothers seized power in northwest Russia. Rurik founds Novgorod (Holmgard).

880 Oleg (Helge) the Wise of Novgorod conquers Smolensk and Kiev. Thus, he controls all trade between Constantinople and the Baltic. Kiev (Great Svitjod) becomes the capital of Russia (kingdom of the **Rus**).

907–911 Oleg leads a fleet against Constantinople (Miklagard), resulting in a trade agreement.

944 Igor of Novgorod forces a new trade agreement on the Byzantine Empire. In return he promises to open Russia (Kievan Rus) to Christian missionaries.

965–967 Sviatoslav crushes the Volga Bulgars and Khazars around the mouth of the Don.

967–971 Sviatoslav attempts to conquer the Donau Bulgars, but in 971 Emperor John I Zimisces drives the Russians from the Balkans.

980 Vladimir the Great wins the power struggle with his brothers in 972. He marries the Byzantine emperor's sister and converts to Christianity in 998. Decrees that all idols are to be burned, and orders all his subjects to let themselves be baptized. Centralization of the kingdom's administration is begun.

1018 Boleslav Chrobry (= the Brave) of Poland attacks Kiev, but is beaten back.

1034–1054 Yaroslav the Wise is absolute monarch in the Grand Duchy of Kiev. He makes the city of Kiev a **metropolis**, subject only to the Patriarchy in Constantinople, organizes the church, and

71. THE GRAND DUCHY OF KIEV 1054

- The duchy's greatest extent, 1054
- → Russian advances in the 10th C.
- → Nomadic incursions in the 11th C.
- → Polish attack on Kiev led by Boleslav the Brave in 1018
- ★ Battlefield

Nursing in a cloister in the 1300s. From a French miniature.

An English woman in a linen scarf is milking. 12th century miniature.

The Fall of Man a depicted in Frenc manuscript from 1200s.

Orkneys
(Scottish from 1469)
Kirkwall (Kirkjuvåg)

KGD. OF NORWAY

Stavanger

Lewis
(Ljodhus) **Caithness**
KGD. OF
Skye (Skie) • Inverness
SCOTLAND
Perth • Aberdeen
Bannockburn • St. Andrews
1314 Edinburgh
Glasgow • Dunbar 1296
Largs
1263
NORTHUMBERLAND
Area claimed by
Scotland 1139-57
ULSTER
Armagh Carlisle • Durham
Downpatrick • **Man** **CUMBERLAND**
IRELAND
Tuam 1171 Lancaster **KGD. OF**
Athlone • Dublin • York
Limerick Anglesey Conway **YORKSHIRE**
Cashel • Waterford **WALES** Nottingham Lincoln
Cork • Cardigan **ENGLAND** **NORFOLK**
St. David's Coventry • Norwich
Llandaff • Bath Evesham Ely **SUFFOLK**
SOMERSET 1265 Cambridge **ESSEX**
Exeter • **DORSET** Oxford London
CORNWALL Windsor Canterbury
Southampton • Arundel **SUSSEX** L'Écluse
Portsmouth Lewes Calais 1348
THE ENGLISH CHANNEL 1264 **ARTOIS** **FLANDERS**
Crecy Courtrai
Barfleur 1346 Bouvines 1302
Bayeux Amiens 1214
Brest Rouen **VERMANDOIS** Soissons
St. Malo • **NORMANDY** Château Reims
Évreux Gaillard 1204 Châlons
Rennes Chartres **ÎLE DE** Marne
La Roche-aux- **FRANCE** Troyes
Moines 1214 **BLOIS** Orléans Sens **CHAMPAGNE**
Nantes Loire Tours **KGD. OF**
ANJOU **FRANCE** Dijon
Bourges **NEVERS**
POITOU Poitiers **BERRY**
1356
72. ENGLAND AND FRANCE 1154-1328 La Rochelle **FRANCHE-**
Taillebourg 1242 Clamont **BURGUNDY** **COMTE**
Saintes **BOURBON** Mâcon Genève
Limoges Lyon
Périgueux Vienne Torino
Bordeaux **AUVERGNE** Valence
AQUITAINE Dordogne **DAUPHINÉ** Genova
Cahors Embrun
GASCOGNE Nîmes Avignon
Albi **PROVENCE**
ARMAGNAC **LANGUEDOC** Arles
Toulouse Béziers Marseille Nizza
BÉARN Carcassonne Narbonne Toulon

NORTH SEA

KGD. OF DENMARK
Viborg • Århus
Ribe • Odense
Schleswig

Hamburg
Bremen
FRIESLAND
HOLLAND Utrecht Münster
Antwerpen **THE HOLY**
Cologne
BRABANT Maas **The Rhine**
Mainz
Trier
Speyer Metz **ROMAN**
Strasbourg
Langres Basel
Besançon

EMPIRE

MEDITERRANEAN

Scottish border in 1157
Eastern limit of Henry II's French possessions in 1154
Eastern limit of English possessions in France a century later
Norwegian possessions until 1266-1469, thereafter Scottish
French royal lands in 1180
France at death of Charles IV in 1328
Ruled directly by King of France
English rule
Other French rulers

ATLANTIC OCEAN

France 1154–1314

1154 Count Henry of Anjou becomes king of England (Henry II), and Duke of Normandy.
1202–1206 Philip II August takes all of the English fiefdoms, except Gascogne.
1294–1304 War with England over Gascogne and Flanders.
1301–1303 Philip IV wins his struggle with Pope Bonifatius VIII.
1312–1314 Bonifatius VIII is forced by Philip IV to dissolve the Knights Templars in 1312.

Knights Templars

In 1119 a clerical order of knighthood was established in Jerusalem to defend the Holy Land. The name Templars comes from the fact that they had their headquarters on the supposed site of Solomon's temple. By around 1300 the order had approximately 20 000 knights and had acquired a great deal of worldly power. The order came to an end when Jacques de Molay, and his next in command, were burned at the stake as heretics in Paris in 1314 (see section from a contemporary miniature **below**).

Francis of Assisi

Francis was born in the village of Assisi in 1182. His father was a rich merchant, and Francis made the most of his position. He dreamed of becoming a knight and troubadour, but at the age of twenty-five he underwent a spiritual crisis, deciding to dedicate his life to the service of God, through acts of charity to the poor and sick. **The picture at the left** is from Giotto's painting: «Francis preaches to the birds».

England 1106–1328

1106 Henry I Beauclerc conquers Normandy.
1154–1189 Henry II strengthens the monarchy and reforms the legal system. The king, who has enormous possessions in France (map nr. 72), begins the conquest of Ireland in 1171.
1215 John Lackland issues the Magna Carta, securing the rights of the barons.
1284 Edward I conquers Wales and lays claim to Scotland.
1314 Robert I (Bruce) drives the English from Scotland following the Battle of Bannockburn.
1328 Of the French fiefdoms, only Britanny and Aquitaine remain English.

Harvesting grain with a sickle. From German 12th century miniature.

Mistreatment of prisoners. Miniature from an English manuscript.

German church, built in 800s and modeled on the church in Jerusalem.

Left: In the beginning of the 15th century Spinello Aretino, with the help of his son, painted sixteen large frescos in the Palazzo Publico (town hall) in Siena. His assignment was to depict the Lombardian cities' struggle against Frederick Barbarossa in the 1170s. The detail shown here shows Italian warriors aboard a galley. This type of vessel was already in use in Roman times. It was normally rowed, but in a strong wind it might also carry one or more sails. The galley was in use throughout the Middle Ages, especially in the Mediterranean, but also in the Baltic.

The German Empire 1024–1250

1024–1039 Conrad II, Holy Roman Emperor from 1027, grants his vassals hereditary rights to their fiefs.

1056–1106 Henry IV, emperor from 1084, employs the gentry and the middle class in his administration instead of the nobility proper. From 1076 he is fighting with Pope Gregory VII over the right to install bishops – the so-called Investiture Struggle. At the Synod of Worms in 1076, Henry is excommunicated. Begs the pope to lift the ban. This is done, but when Pope Gregory again excommunicates him in 1080, Henry drives the pope from Rome. Compromise reached.

1125–1137 Lothar II of Saxony, emperor from 1133, carries out a campaign of expansion north to the Baltic.

1152–1190 Frederick I Barbarossa becomes emperor in 1155. The northern Italian city-states (Lombardian League) defeat Frederick Barbarossa at Legnano in 1176, but the emperor manages to put down the rebellion. Frederick Barbarossa drowns while on the way to the Holy Land in 1190 (the Third Crusade).

1194–1250 Frederick II (see p. 50) rules nearly all of Italy, treats Germany as a dependency and establishes a modern, unified state in southern Italy. In 1227 Frederick II too is at odds with the pope, because the emperor has broken off a crusade. Frederick is excommunicated, but a year later he goes on a crusade anyway, and in 1230 the pope lifts his ban.

Map nr. 73 shows the Holy Roman Empire at the time of Frederick's death in 1250.

73. HOHENSTAUFFEN POWER IN EUROPE
— Boundary of the Holy Roman Empire at the death of Frederick II in 1250

Two-wheeled English cart. Miniature in a 14th century manuscript.

Wine grapes being picked. French relief from beginning of the 1200s.

Grapes being trampled. French relief from beginning of the 1200s.

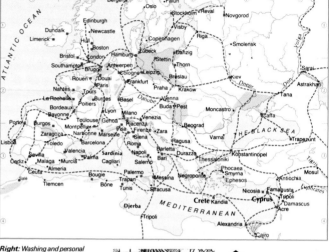

Above: Genoa's harbour at the end of the 1400s. Section from a contemporary woodcut. Genoa was just as wealthy and powerful as Venice and had hegemony at sea over an equally large area, but the city never experienced the same cultural blossoming as Venice and other Italian economic centres. The wealthy traders of Genoa preferred money to culture.

Below: A 15th century woman about to give birth. At left bath water is prepared for the newborn. Thereafter the child will be swaddled, and have its legs wrapped, so that they will grow straight.

Right: Washing and personal hygiene were not given high priority in the Middle Ages. That is **one** reason why the great epidemics could spread so quickly. In addition, a general lack of cleanliness led to an obtrusive stench of body odour. Now and then – and in any case at Christmas – people would climb into the tub, as in the scene shown here, where a man and wife seem to be enjoying their bath. From a German woodcut from the early 1500s. In the following centuries, members of the upper classes attempted to make body odour less noticeable by the generous application of perfumes.

74. INDUSTRY AND TRADE IN 13TH CENTURY EUROPE

Major production areas for:
- Grain
- Wine
- Olive oil
- Important trading centres and markets
- --- Important trade routes

75. SPREAD OF THE BLACK DEATH, MID-14TH CENTURY

Areas hit by the plague in:

1346   1347   1348

1349   1350   1351-53

Areas with little or no plague

The roasting of fish. English miniature from around 1330.

Plague doctor in prescribed garb. «Beak» on mask was to stop contagion.

Tilling the soil with a harrow. English miniature from around 1340.

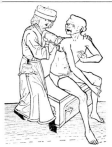

The Black Death

Ca. 1345 rumours spread of epidemics killing people by the hundreds of thousands in the East. In the West people at first took a certain satisfaction in the knowledge that the unbelievers were being punished. Not until 1347, when some sailors in Genoa had been infected, did they see that the sickness was not discriminating between non-believers and Christians.

Bubonic plague is caused by a bacillus in the blood of certain animals, especially rats. Humans were infected by fleas that infested both animal and human hosts, or through direct human contact. **The woodcut above** shows a physician lancing a bubo, or boil, unaware that countless bacteria were carried further in the many folds of his clothing.

In Paris there were eight hundred deaths per day. On the whole, cities were the hardest hit. As much as a third of the population died in areas affected by the plague (map nr. 75).

76. EUROPE CA. 1400
- ☐ Genoa and possessions
- ■ Venice and possessions

Venice and Genoa

Throughout the 14th century Venice and Genoa struggled for control of the Mediterranean. The two city-states had each established strategic outposts, both at home, and in the Greek areas near the Black Sea (map nr. 76).

The entrance to the Black Sea was of special importance. Venice had acquired a trade monopoly in Constantinople, while Genoa held the fort of Pera north of the city.

Switzerland 1315–1499

1315 The original cantons of Uri, Schwyz and Unterwalden, which since 1291 have been united in the Swiss Confederation, having sworn 'Eternal Union', defeat Duke Leopold of Austria at the Battle of Morgarten. The other Swiss cantons then join, one after the other, the Eternal Union.

1389 Peace with the Hapsburgs, who time and again have attempted unsuccessfully to subjugate the cantons.

1499 Maximilian I recognizes the Swiss Confederation's political independence at the Peace of Basle. Switzerland remains only formally a part of the German Empire.

77. THE SWISS CONFEDERATION 1315–1536
- The original cantons
- Confederation 1481
- Confederation 1536
- Protected by the confederation
- Subject to one or more cantons ◀

The plague reached Crimea from the area around Lake Balkhash

English soldier with longbow at Battle of Azincourt in 1415.

Style-conscious young man with doublet and long stockings. 1400s.

Threshing grain with a swingle. From a calendar page from the 1400s.

The Hundred Years' War between England and France

The war actually lasted 116 years (1337–1453). At the outbreak there was no clear distinction between the two countries. The upper classes spoke French, and the Channel was at that time not a natural border.

The greatest victory for the English in the Hundred Years' War was won at Azincourt in 1415. The highpoint for the French was Joan of Arc's rescue of Orléans in the summer of 1429 (see column 4).

The Hanseatic League

was actually formed by a «hansa» – a German merchant guild – in Wisby on the Swedish island of Gotland. Through the years, however, the organization strengthened its position and economic power, so that by around 1400, it controlled all trade in the North Sea, Skagerrak, Kattegatt and the Baltic (map nr. 79). Hamburg and Lübeck were the Hansa traders most important towns. **The picture above** – from a miniature in Hamburg's municipal laws from 1497 – shows lively traffic in Hamburg's harbour.

The Maid of Orléans

In the summer of 1429, seventeen-year-old Jeanne d'Arc led a division of French soldiers from one victory to the next. It wasn't until May of the following year that the English managed to capture her. She was put on trial as a heretic and witch, and on May 30, 1431 the legendary maiden was burned at the stake in Rouen. The **picture below** shows her being tied to the stake.

		The Black Prince 1356		French fiefs
		John the Good 1356		Fiefs subject to English crown
		Edward III 1359-60		
		Eastern boundary of fief subject to English crown		Burgundian fiefs
		Henry V 1415		Boundary of the Kgd. of France
		Joan of Arc's ride to Reims, Summer 1429		

78. ENGLAND AND FRANCE IN THE HUNDRED YEAR'S WAR

Town hall – Palazzo Vecchio – in Firenze, built at the end of the 13th century.

A generous amount of wine is drunk at the inn. Italian miniature from the 1300s.

Lady of the castle hauls in her lover German miniature from the 1300s.

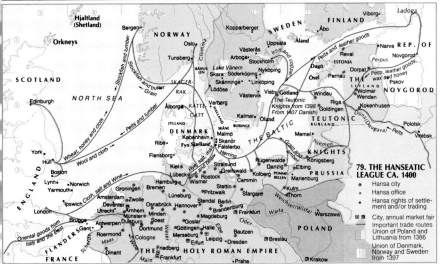

79. THE HANSEATIC LEAGUE CA. 1400
- Hansa city
- Hansa office
- Hansa rights of settlement and/or trading
- City, annual market fair
- Important trade routes
- Union of Poland and Lithuania from 1386
- Union of Denmark, Norway and Sweden from 1397

Italy 1250–1454

1250 At the death of Frederick II, the Hohenstaufens rule almost the entire peninsula (map nr. 73, p. 53).

Around 1300 The northern Italian city-states gain independence, as bishops and imperial civil servants are pushed aside.

A cultural golden age:
Dante Alieghieri (1265–1321)
Francesco Petrarca (1304–1374)
Giovanni Boccaccio (1313–1375)
Giotto di Bondone (approx. 1266–1337)

Around 1450 Florence (Firenze), ruled by the Medici family, becomes the financial centre of Europe. It is the time of the Early Renaissance.

In **1454**, at the so-called Treaty of Lodi, Cosimo de' Medici the Elder, and Francesco Sforza, succeed in bringing about peace in Italy (map nr. 80).

Below: In a fresco in Santa Trinita in Florence, painted by Dominico Ghirlandaio, the mightiest of all the Medicis – Lorenzo de' Medici il Magnifico (1449–1492) – is placed in the foreground is probably Lorenzo's son, Giovanni, who would later become Pope Leo X.

80. RENAISSANCE ITALY AFTER THE PEACE OF LODI 1454

Lecture in progress at a German university around 1500.

A carrier pigeon being sent off. 15th century woodcut.

Emperor Charles V. presiding over the Parliament in Worms in 1521.

81. SPREAD OF THE REFORMATION TO ABOUT 1550

- Protestantism established by 1550
- Catholic majority, reformatory advances
- Catholic majority never under threat
- ▬ Universities and colleges under strong reformatory influence
- 1537 Date when protestantism officially adopted

Boundaries refer to the situation in 1589, see also map 96

ses the patron saint of miners, St.Anna, that he will become a monk if she spares his life.
1507 Luther, now a monk, is ordained in the cloister of the Augustinian hermits in Erfurt.
1508 Transfers to the monastery in Wittenberg.
1517 Angered at the Dominican monk Johan Tetzel's tasteless peddling of indulgences in Saxony, Luther, on October 31, tacks his ninety-five theses against the traffic in indulgences to the door of the Castle Church of Wittenberg.
1520 Pope Leo X (Giovanni de' Medici) excommunicates Luther, but the papal bull threatening excommunication is thrown on a bonfire in Wittenberg. In the same year Luther publishes several programmatic essays.

Below: Alter piece by Lucas Cranac the Elder, painted in 1547, the year after Luther's death. Between the reformer in the pulpit, and the congregation, hangs the crucified Christ – according to Luther, man's only guarantee of salvation.

The Reformer Martin Luther

1483 Mine worker Hans Luther and Margarethe Ziegler have a son, and name him Martin.
1484–1501 The family moves from the town of Martin's birth, first to Magdeburg and then to Eisenach. Martin is strictly disciplined by his mother and father, and school is a place of rote learning and whippings.
1501 Martin is sent to Erfurt to study law.
1505 Student Martin Luther experiences a tremendous thunder storm during a journey and promi-

Dutch bride and groom from 1434. From painting by Jan van Eyck.

Two chastity belts. One on left from the 1400s, the other from the 1500s.

A gentleman negotiating with a moneylender. German woodcut from 1486.

Above: Prince John Frederick of Saxony, and leading Reformation figures. From yet another painting by Lucas Cranach the Elder from around 1530. Luther is seen standing at left. On the other side of the prince stands the Swiss reformer Ulrich Zwingli (1484–1531), and farthest to the right we see Luther's melancholy friend, Philip Melanchton.

1521 Professor Luther refuses to retract his earlier statements. He is made an outlaw, but by now he is already under the protection of Prince Frederick of Saxony at Wartburg. There, under the name of Junker Jürgen, Luther works on his translation of the New Testament.

1525 Marries former nun Katharine von Bora, who bears him six children.

1529 The Diet of Speyer reaffirms the judgment of 1521, but a minority lodges a formal protest. They are hereafter referred to as *protestants*.

1530 Luther's good friend, Philip Melanchton (see illust. above) formulates the «Augsburg Confession». In December of the same year the protestants form the Schmalkaldic League. Its purpose is to defend the teachings of Luther against the emperor by force of arms.

1546 Martin Luther dies on February 18 and is buried in the cemetery of the Castle Church in Wittenberg. Reformation established in Northern Europe.

82. THE SCHISM IN THE CATHOLIC CHURCH 1378-1417

▨ Adherents of Rome
▨ Adherents of Avignon
▨ Loyalties divided

Boundaries refer to the situation in ca. 1400

Left: The movement led by the Frenchman Jean Calvin (1509–1564) opposed anything that distracted the worship service. This section of a painting from 1564 depicts the interior of a church in Lyon, furnished in a style both Calvinist and Spartan, in which the pulpit, not the alter, is central.

83. THE CHURCH IN EUROPE CA. 1500

✛ Archbishopric
▨ Under papal administration

▨ The Ottoman Empire
Other colours used to depict archbishoprics

Sailing ship from the 1400s, heavy and stabile, with all sails set.

Arab slaves toiling for European colonizers

A 16 cm high woo sculpture from the Congo (Zaire).

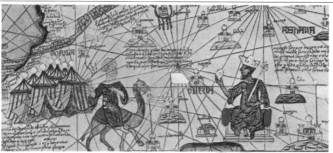

Left: Even before colonization began in earnest, foreign merchants had contacts with Africa. Ships from Spain, Portugal and Italy made regular stops at African ports to barter with the natives. Horses, cloth and implements were traded for dates, olives, cola nuts, cotton, copper, and, above all – gold. Depicted in this section of a Catalonian map from 1375, an Arab trader on a camel is approaching the king of Mali to conclude a bargain. The king is holding a clump of gold in his hand.

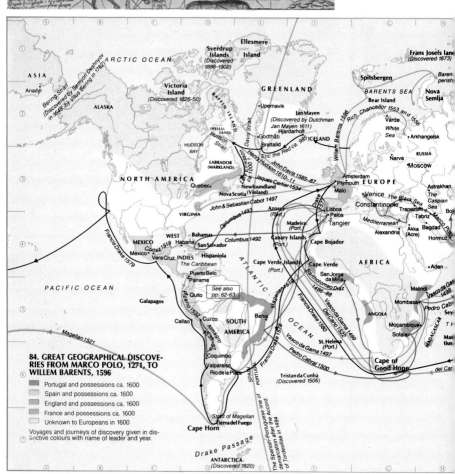

84. GREAT GEOGRAPHICAL DISCOVERIES FROM MARCO POLO, 1271, TO WILLEM BARENTS, 1596

- Portugal and possessions ca. 1600
- Spain and possessions ca. 1600
- England and possessions ca. 1600
- France and possessions ca. 1600
- Unknown to Europeans in 1600

Voyages and journeys of discovery given in distinctive colours with name of leader and year.

Spaniard Diego de Almagro, who discovered and conquered Peru.

Kangaroo hunt in Australia. Cliff painting from Kimberley.

Spanish conquistador beating an «uppity» indian in Mexico.

Explorations and Discoveries 1271–1596

1271–95 Marco Polo sets out from Venice and explores China.

1486–88 Portuguese captain **Bartholomeo Diaz** discovers the sea route around the southern tip of Africa.

1492–93 Christopher Columbus (illus. right) reaches the Lesser Antilles in Central America and names the island San Salvador.

1497 English-Italian explorer **John Cabot** and son **Sebastian** reach Virginia, which they claim for England.

1497–99 Portuguese **Vasco da Gama** (illus. right) finds the sea route to India.

1500 Portuguese **Pedro Alvares Cabral** reaches the land that will one day become Brazil, but hurries on to Portuguese East Africa and India.

1519 Spaniard **Fernando (Hernando) Cortez** leaves Cuba, which he helped to conquer in 1511–12, and conquers Mexico, defeating king Montezuma II (see also map nr. 87, p. 63).

1519–22 Portuguese **Fernando Magellan** is the first to circumnavigate the globe. In 1521 he is killed on the island of Mactan in the Philippines. **Juan Sebastian del Cano** sails his vessel home. **1531–37** The uncouth and brutal Spaniard **Francisco Pizarro** conquers Cuzco and the Inca Empire. His companion, **Diego de Almagro**, crosses the Andes, where he discovers and conquers Chile. **1534–42** Frenchman **Jacques Cartier** undertakes three journeys to North America, primarily in search of a new route to India.

1553 and 1556 Francis Drake completes the world's second circumnavigation (map nr. 84).

1585–87 Englishman **John Davis** explores the coasts of Greenland, Baffin Island and Labrador.

1594 and 1596 Dutchman **Willem Barents** discovers Nova Zemlya, Spitsbergen and Bear Island.

Top right: Christopher Columbus (1451–1506), son of a weaver from Genoa, who, without knowing it himself, discovered America. From a painting in Museo Curco in Como.
Middle: Vasco da Gama (ca. 1460–1524), left Lisbon on July 8, 1497 for India. On May 20, 1498 he landed in Calcutta, and in July of 1499 he was back in Lisbon. In 1524, shortly before his death, he was made viceroy of India.
Bottom: Marco Polo (ca. 1255 – ca. 1325), his father and uncle are received Kublai Khan in Khanbalik (Beijing) in 1275. The journey to China had taken 3½ years, and Marco remained 17 years in the service of the Khan.

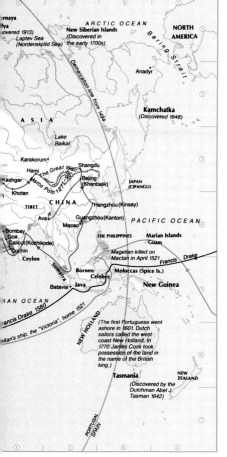

Reconstruction of a birch bark canoe from Canada.

French courtesans on their way to New Orleans by order of Louis XIV.

Tent ('wig-wam') used by the American Plains Indians.

Left: The terraced pyramid of Chichén Itza was erected during the time of the so-called Northern Mayan Culture on the Yucatán Peninsula (map nr. 87). Stairways lead up to a temple at the top.

North and South America 1620–1775

North America

1620 «The Pilgrim Fathers» arrive in Massachusetts on the «Mayflower». British colonization is accelerated. The Spanish have been established since 1535 in the viceroyalty of New Spain (map nr. 85).

1626 New Amsterdam (New York) is founded as the capital of the Dutch possessions.

1664 England conquers the Dutch colonies and occupies New Amsterdam. **1756–63** The British-French Colonial War over control of the North American territories. England wins, and at the Peace of Paris in 1763, all of North America became Anglo-Saxon (map nr. 85).

1774 Representatives of «The Thirteen Colonies» meet. The First Continental Congress adopts a Charter of Rights and votes to continue the boycott of English goods.

1775 The American War of Independence (see pp. 92–93).

South America

1650 Spain has control in two viceroyalties: Peru and La Plata. The Portuguese have occupied large areas on the east coast.

1775 Both Spain and Portugal have expanded their possessions in the interior of the continent.

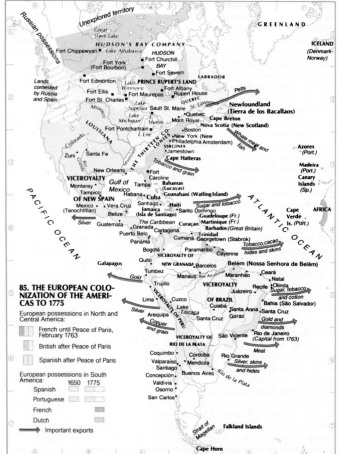

85. THE EUROPEAN COLONIZATION OF THE AMERICAS TO 1775

European possessions in North and Central America:

- French until Peace of Paris, February 1763
- British after Peace of Paris
- Spanish after Peace of Paris

European possessions in South America:

	1650	1775
Spanish		
Portuguese		
French		
Dutch		

→ Important exports

The Spanishs tortured the Incas as well. Here one of them has his eyes gouged out.

Two Aztec chieftains surrender to Hernán Cortez.

Aztec woman weaving with loom made fast to a tree.

86. THE INCA EMPIRE

- Chimu Kgd. before 1470
- Empire's greatest extent 1530
- --- Regional border
- — Inca highways
- Some modern names given for purposes of orientation

Bottom of the previous page: Mayan warrior from the 700s armed with some imaginative weapons. Until a group of researchers found pictures like this – in a ceremonial building in Bonampák in the state of Chiapas in 1946 – the Mayans were thought by most to have been an almost pacifistic people.

Above: The Incan capital of Cuzco as sketched by a Spanish artist in the 1600s. Spanish/European influence has already made itself felt, not least in the Sun Temple, Coricancha, seen at the left. The plan of the city itself, however, with its straight streets, is from the Inca period.

87. AZTEC AND MAYAN CIVILIZATION

- Aztec heartland in 1486
- Aztec Kingdom ca. 1520
- Southern Mayan Civilization ca. 300-900
- Northern Mayan Civilization ca. 900-1200
- Mayan mountain kingdoms ca. 1500
- → Cortéz' route 1519-20
- Present-day boundaries ▼

Below: Toltec warrior from the temple in Tula in present-day Mexico. Originally this and others like it functioned as supporting columns for the temple's roof construction. They are carved out of basalt and are over 4.5 m high.

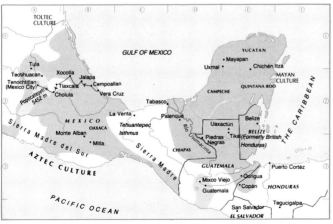

The mausoleum of sultan Ghiyas-udin Tughluq, completed in 1325.

A European is carried in a sedan chair by Indian slaves.

A European oversee the harvesting of pepper on the coast of Malabar.

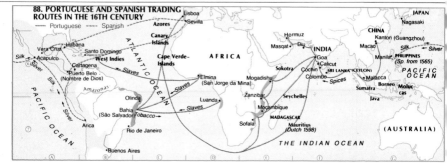

88. PORTUGUESE AND SPANISH TRADING ROUTES IN THE 16TH CENTURY

— Portuguese   ---- Spanish

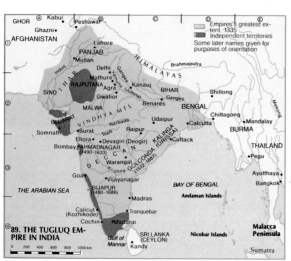

89. THE TUGLUQ EMPIRE IN INDIA

Empires's greatest extent. 1335
Independent territories
Some later names given for purposes of orientation

0 200 400 600 800 1000 km

India in the 14th Century

Sultan Ghiyas-ud-din Tughluq founds the Tughluq Dynasty in 1320, which retains power in India for almost eighty years. Under his son, Mohammed bin Tughluq (1325–51) the kingdom sees its greatest expansion (map nr. 89). His nephew, Firuz Shah cannot hold the empire together; and from 1398 it begins to dissolve.

Above: The Mogul emperor Akbar the Great, who governed most of northern India 1556–1605, was an active participant in the cultural life of the kingdom, and showed great tolerance for other religions. This section of a contemporary miniature shows him in conversation with a Jesuit priest (in black) and Muslim philosophers. Akbar claimed that the various religions in reality did not differ from one another beyond the point of using different names for one and the same almighty, monotheistic divinity.

Left: Portuguese colonizer Affonso d'Albuquerque (1453–1515) sailed to the East in 1503, and 6 years later was made governor general of the Portuguese possessions in India, with headquarters in Goa on the west coast.

90. THE MAJAPAHIT EMPIRE. DUTCH CONQUESTS IN INDONESIA

Dutch conquests:

Empire ca. 1400   17th C.   18th C.

Goa ca 1500 – Indian capital – centre of trade between Asia and Europe.

An Indian prince visits one of his courtesans. Both are smoking hookah pipe

Indian servant holds a parasol over a European merchant in Goa.

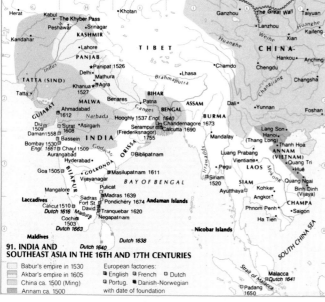

91. INDIA AND SOUTHEAST ASIA IN THE 16TH AND 17TH CENTURIES

▨	Babur's empire in 1530
▨	Akbar's empire in 1605
▨	China ca. 1500 (Ming)
▨	Annam ca. 1500

European factories:
■ English ■ French □ Dutch
□ Portug. ▪ Danish-Norwegian
with date of foundation

Top right: Detail from a bronze relief from Benin. The inhabitants of the African kingdom of Benin in present-day Nigeria (map nr. 92) developed, beginning about 1300, a unique method for casting bronze and brass.

India 1503–1707

1503–19 Portuguese establish trading stations on the west coast and on the island of Ceylon.

1526 Mongol prince Babur, a descendent of Timur-lenk, founds the Mogul Empire in India after victory in the Battle of Panipat, where his 12 000 men, armed with artillery, defeat an army of 100 000.

1527 Babur consolidates his position after the Battle of Khanua where he defeats the Rajput princes' enormous forces.

1556–1605 Under Akbar the Great (illus. p. 64) India has its greatest period of expansion.

1608 The English East-India Co. opens a trading post in Surat.

1658–1707 The Great Mogul Aurangzeb, a devoted follower of Islam, attempts through endless warring to put an end to Hinduism. After his death, the empire begins to unravel.

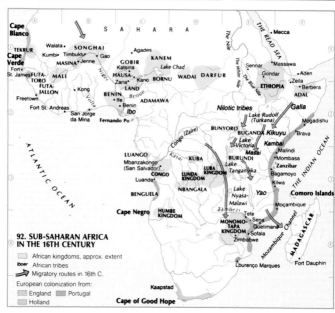

92. SUB-SAHARAN AFRICA IN THE 16TH CENTURY

▨	African kingdoms, approx. extent
Iboer	African tribes
⇗	Migratory routes in 16th C.

European colonization from:
▨ England ▨ Portugal
▨ Holland

Solider with musket around 1600. To aim the gun was laid in the cradle (left).

Musketeer fills the barrel of his musket with powder from a powder horn.

Goods being transported by sled on a winter's day in Amsterdam ca. 1600.

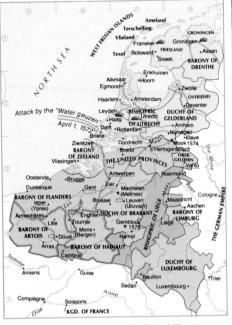

93). The Duke of Alba resigns as governor in 1573. Leyden, under seige by the Spanish, is liberated by William of Orange in 1574.

93. THE DUTCH REVOLT 1559-1648

↗ Spanish campaign in the North 1572-74

— Boundary of the Catholic Union of Arras 6 January 1579

The United Provinces after Union of Utrecht 23 Jan. 1579

Gent Towns that briefly joined the United Provinces

Became part of United Provinces in 1648

The Spanish Netherlands:

▨ in 1579 ▨ in 1648

▨ Church lands

⚘ Protestant } universities
⚘ Catholic

The Netherlands 1559–1648

1559 The Spanish king Philip II leaves the country, making his half-sister, Margaret of Parma, governor. She continues the persecution of the protestants.

1566–67 Protestant riots due to famine. Philip II sends the Duke of Alba with an elite force of 20 000 to the provinces. Courts of the Inquisition are established, and mass executions are carried out.

1573–74 Spanish campaigns against rebels in the north (map nr.

1579–81 Provinces in the north form the Union of Utrecht in 1579; later The United Netherlands. An assembly of the seven estates deposes Philip II in 1581.

1648 At the Peace of Münster, Spain recognizes the United Netherlands. Estates General (map nr. 93) united with the northern provinces. Spanish Netherlands (Belgium) remain under the rule of the Hapsburgs.

The Great Fire in London in 1666

Londoners had scarcely begun to recover from a terrible outbreak of bubonic plague that claimed 70 000 lives in 1665, when a massive fire broke out on Sept. 2, 1666.

It started in a street in City where warehouses were used to store jute and tar. The fire moved quickly through the wooden structures. The Stock Exchange, St. Paul's Cathedral, eighty-eight other churches and some 13 000 houses burned in 5 days. Some 100 000 people were left homeless.

Above: Children at play in a Dutch town. From a painting by Pieter Brueghel the Elder (approx. 1525–69). The children are turning cartwheels and playing leapfrog.

94. LONDON IN SHAKESPEARE'S AND PEPYS' TIME

▨ Extent ca. 1600 (200 000 inh.)

— London Wall

⋯ Extent of the Great Fire 1666

London Bridge was the capital's only one across the Thames until Westminster Bridge was built in 1750

Various forms of insult might lead to a duel.

Prince James (later James II) playing tennis around 1625.

Sir Walter Raleigh, who brought back tabacco from the Americas.

Above: Queen Elizabeth I of England as she is portrayed in the so-called Armada Portrait from the end of the 1530s, i.e., after the Spaniards'«Invincible Armada» (top left) had been crushed in 1533. The Virgin Queen was the daughter of Henry VIII and Anne Boleyn, who was accused of infidelity and executed at the age of twenty-nine. Elizabeth was herself twenty-five when she ascended England's throne, and when she died in 1603 she had ruled her country for nearly forty-five years and lent her name to an entire epoch in European history.

England and Scotland 1553–1673

1553–58 Struggles between religious factions, and an attempt to reintroduce Catholicism during the reign of Mary Tudor (Bloody Mary). England becomes involved on the side of Spain in the war with France.

1558–1603 Under Elizabeth I both the monarchy and the Anglican church see greater stability. In 1584 Walter Raleigh is given the task of claiming new lands in America. He calls the coastal region north of Florida «Virginia» – after the «Virgin Queen» – and sends colonists to the area. The execution of Scottish-French queen Mary Stuart in

1587 leads to war with Spain. The «Invincible Spanish Armada» is annihilated in 1588. The East India Company is established in 1600.

1603–25 James VI, Mary Stuart's son with Lord Darnley and King of the Scots from 1567, is from 1603 also King of England, taking the name of James I. He persecutes the Puritans and, following a Catholic assassination attempt (The Gunpowder Plot) in 1605, also the Catholics. Religious minorities begin emigrating to North America.

1625 James I's son, Charles I, becomes king. He governs without a parliament from 1629 to 1640

because of a dispute about taxation. Convocation of *The Long Parliament* leads to civil war in 1642.

1642–49 Charles I is taken captive by the Parliamentary Army. Oliver Cromwell defeats the Catholics (the Scots) at Preston in 1648 (map nr. 95). He removes all of his opponents in the Parliament, and has Charles I executed in 1649.

1653–58 Oliver Cromwell governs without parliament, as absolute ruler with the title of *Lord Protector*. War with Spain 1654–58.

1660–73 Charles I's son, Charles II, is elected king. London burns in 1666 (see p. 66). To counteract Charles II's Catholic sympathies,

95. THE BRITISH ISLES IN THE MID-17TH CENTURY

Civil War between King (K.) and Parliament (P.) 1642-49

▢ Controlled by P. end of 1643
▢ Controlled by K. end of 1643
◗ Under P.'s control 1645
➤ Cromwell's campaign 1649-51
▢ Reserved for Cromwell's protestant veterans
▢ Reserved for banished Irish
★ Major battlefield with date

the Test Act is passed, requiring all civil servants to receive Holy Communion in accordance with the Anglican rite.

96. THE RELIGIOUS POSITION IN EUROPE CA. 1560

▢ Roman catholics   ▢ Hussites
▢ Orthodox cathol.   ▢ Anglicans
▢ Lutherans   ▢ Muslims
▢ Calvinists

Two soliders have laid their muskets in the cradle and are opening fire.

Here a fierce sword fight is underway.

97. SPAIN'S EURO-PEAN EMPIRE 1580

Spain's possessions
—— Bound. Roman Emp.
— — The Spanish Armada 1588

IRELAND
KGD. OF ENGLAND
Fotheringhay
Stratford • Cambridge • Leyden • Amsterdam • Osnabrück
WALES • Oxford • London • Münster
Southampton • Dover
Plymouth • Calais • FLANDERS • THE NETHERL.
Cateau-Cambresis • Brussel • LUXEM-BOURG
Amiens • Guise • THE HOLY ROMAN
Beauvais • BOHMEN
Caen • NORMANDIE • Reims • Vass. • SACHSEN
Alençon • Paris • Vassy • Nürnberg • EMPIRE • Vienna • KGD. OF
Vendôme • Amboise • Nevers • FRANCHE- • BAYERN • AUSTRIA • HUNGARY
Nantes • ANJOU • COMTE • STEIERMARK
CHAROLLES • THE ALPS • TIROL
La Rochelle • KGD. • DUCHY • REP. OF VENEZIA
OF FRANCE • OF SAVOY • Venezia • THE
Bordeaux • Milano • Genoa • ADRIATIC • OTTOMAN
LANGUEDOC • Avignon • SEA • EMPIRE
La Coruña • Santander • San Sebastian • PROVENCE
PYRENEES • Marseille • Firenze • PAPAL
Braganza • Perpignan • TOSCANA • STATES • KGD. OF
KGD. OF • ROUSSILLON • Piombino • Elba • Roma • Pontecorvo
Escorial • NAVARRA • Corsica • Benevento
Lisboa • KGD. OF • KGD. OF • Lerida • Barcelona • Napoli • NAPOLI
Madrid • CASTILLA • ARAGON • CATALONIA
Sevilla • SPAIN • Valencia • Menorca • Sardinia
Palos • Cordoba • VALENCIA • Mallorca
Cadiz • ANDALUSIA • Granada • Ibiza • Baleric Islands • Palermo
Tanger • GRANADA • KGD. OF
Ceuta • Gibraltar • Cartagena • SICILY
Melilla (Spanish 1496) • MEDITERRANEAN
Oran (Spanish 1509) 1510-19) • Algiers • Bougie • Tunis • Malta
(Spanish 1535-74) • (Spanish • (Spanish 1535-74) • (Knights of St. John)
Bizerte • Bona • 1535)
(Spanish 1535-74) • (Spanish • Kairouan
1510-55)

THE ENGLISH CHANNEL
Ca. 65 ships Sept. 21, 1588
The Grand Armada ca. 130 ships
ATLANTIC OCEAN
Ca. July 14, 1588
May 14, 1588

NORTH SEA
•Hamburg
Berlin • Elbe
Cologne
Rhine
ALSACE
LORRAINE
BRETAGNE

Spain under Philip II 1556–98

In contrast to his father, Charles V, Philip II was a genuine Spaniard, fanatically Catholic, dutiful and industrious. The picture of him – **top of this page** – is from Alonzo Sachez Coello's contemporary portrait. One might easily think this a clergyman or monk, rather than the ruler of the dominant great- power of the day.

Philip II had no real interest in life at court, with grand balls and hunts. His favourite place was sitting behind his desk. Nor was he a warrior, but when Elizabeth I had Mary Stuart executed in 1587, Philip sent «The Great Armada» to crush England. This ended in tragedy for Spain.

98. GERMANY IN THE THIRTY YEAR'S WAR 1618-48

Habsburg Lands:

    Austrian   Spanish
→ Christian IV's campaign to Lutter am Barenberge 1625-26
⇢ Gust. Adolphus' camp. 1630-32
→ Tilly's most important campaigns
--⇢ Wallenstein to Lützen 1632
★ Major battlefield with date
    Church lands
    France and French possessions in Germany
Bavaria, Brandenburg and Saxony are shown in distinctive colours

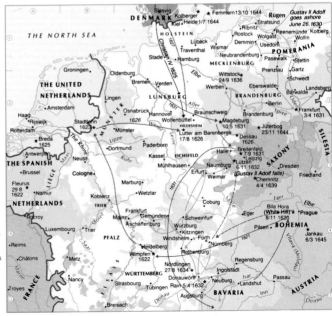

Slesvig • Kolberger • Femmern 13/10 1644 • Rügen • Gustav II Adolf
DENMARK • Kiel • Heide 1/7 1644 • Stralsund • goes ashore
THE NORTH SEA • HOLSTEIN • Ribnitz • June 26.1630
Lübeck • Rostock • Wolgast • Peenemünde • Kolberg
Traventhal • Wismar • Demmin • Usedom • Wollin
Groningen • Stade • Hamburg • Neubrandenburg • POMERANIA
Oldenburg • Elbe • MECKLENBURG • Pasewalk • Stettin
THE UNITED • Bremen • Verden • Wittstock • Prenzlau • Schwedt
NETHERLANDS • Lingen • Aller • 24/9 1636 • Eberswalde • Landsberg
Amsterdam • Osnabrück • Werben • BRANDENBURG • Bärwalde
Haag • 1627 • Hannover • Braunschweig • Berlin • Spree
Rijswijk • Stadtlohn • Wolfenbüttel • Brandenburg • Frankfurt
Rotterdam • 1623 • Münster • HILDESHEIM • Magdeburg • 3/4 1631
Breda • Lippe • Paderborn • Lutter am Barenberge • 10/5 1631 • Juterbog • SILESIA
1625 • Dortmund • 17/8 1626 • Dessau • 23/11 1644
Antwerpen • Ruhr • Kassel • EICHSFELD • Halle • 1626 • Breitenfeld
THE SPANISH • Neuss • Cologne • Mühlhausen • Naumburg • 7.9 1631 • Dresden
Brussel • Marburg • Erfurt • Weimar • Lützen • Gustav II Adolf falls) • Friedland
NETHERLANDS • Wetzlar • 6.11 1632 • Chemnitz • 4/4 1639
Fleurus • Namur • Koblenz • Coburg • Saale • Eger
29.8 • TRIER • Bila Hora • Elbe
1622 • Frankfurt • Schweinfurt • Eger • (White Hill) • Prague
Rocroy • Mainz • Gemünden • 8/11 1620
Luxembourg • Trier • Aschaffenburg • Würzburg • Pilsen • BOHEMIA
Reims • PFALZ • Kitzingen • Nürnberg • 1621 • Jankau
Windsheim • Fürth • 6/3 1645
Châlons • Heidelberg • Rothenburg • Vltava (Moldau)
Metz • Wimpfen • Nördlingen • Regensburg
1622 • 27/8 1634 • Ingolstadt • Passau
WÜRTTEMBERG • Donauwört • Neuburg • Isar • Landshut • AUSTRIA
Troyes • Nancy • Strasbourg • Tübingen • Rain 5/4 1632 • Augsburg • Donau
Breisach • BAVARIA • Inn • Donau

DENMARK • CHRISTIAN IV 1625 • LÜNEBURG • 1631 • SAXONY

One soldier uses his sword, while the other swings his musket.

Here fighting is hand-to-hand.

They even go at each other with their musket cradles. Drawings from 1600s.

Above: *Christian IV of Denmark-Norway. Painting by Peter Isaacsz, 1610. Christian IV was made king at the death of his father in 1588, but did not actually govern the country until his 19th year, in 1596.*

99. SWEDEN IN THE 17TH CENTURY. CHARLES XII'S CAMPAIGNING 1700-1718

- ▦ Sweden in 1560
- ▨ Annexed from Denmark-Norway 1645-60
- ▦ Other conquests 1561-1660
- → Campaign against Copenhagen 1657-58 (Charles X)
- → Charles XII's camp. 1700-1718
- ⇢ Charles' ride to Stralsund 1714
- ★ Major battlefield with date

The Thirty Years' War 1618–1648

The Thirty Years' War, even though there is evidence that it began as early as 1609, ended 1659. This was mainly a struggle between the Hapsburgs and France for supremacy in Europe. Fighting on the side of the Hapsburgs was the Catholic League, supported by Spain. France got support from the Protestant League, incl. Christian IV (see above) of Denmark and Gustavus II Adolphus of Sweden.

Sweden 1617–1718

1617 Peace of Stolbova. Sweden receives Keksholm and Ingermanland (map nr. 99).
1629 Sweden receives Livonia and Riga at the Peace of Altmark.
1630–48 Thirty Years' War. Gustavus Adolphus falls at Lützen (1632).
1643–45 War with Denmark-Norway. Sweden receives Jämtland, Härjedalen, Halland, Gotland and Ösel.
1700–21 Great Northern War. Charles XII falls at Fredriksten (1718).

Right: *Queen Christina of Sweden. From a painting by Sebastien Bourdon, 1653. She was keenly interested in French intellectual life, and as reigning queen from 1644, she wanted to make Sweden a cultural great-power, in keeping with the country's political position. In 1654 Christina abdicated, and on Christmas Eve of the same year she secretly converted to Roman Catholicism. Later she lived mostly in Rome, and was buried in The Church of St. Peter in the Vatican in 1689.*

Dispatch rider
brings news of the
Peace of Westphalia
in 1648.

Child in a walker.
Drawn from an en-
graving from 1636.

Print shop in 1568.
At right the type is
being inked, at left
a finished sheet.

The Peace of Westphalia 1648

Politically, the conditions of the peace of 1648 were disastrous for the German Empire. Of the in all 234 states and 51 cities that had been subject to the emperor, most were now granted sovereignty. The emperor's influence after 1648 was based solely on his position as ruler of Bohemia, Hungary and the Hapsburgs' hereditary lands in Austria. Of the sovereign states, Brandenburg became one of the most powerful (see illust. below).

The most important clause, from a religious viewpoint, was that which upheld the right of the governments themselves to decide religious matters. The distribution of Catholicism and Protestantism in the German areas was, therefore, roughly that of the present day.

The relative positions of Sweden and France too were strengthened by the War. Both received territory previously held by the German emperor. France got portions of the Alsace. Sweden was given Hither Pomerania, with Stettin, Wismar and Ren, and the Duchy of Bremen and Verden. Both countries were also given status as «protectors of the peace», giving them the formal right to intervene in German internal affairs.

Below: In 1618 East Prussia was united through succession with Brandenburg, and from 1648 Further Pomerania was also part of the kingdom.
Brandenburg-Prussia's leader Frederick Wilhelm is seen here accepting the homage of the Prussian Estates on October 18, 1663. In the background at the right is the Castle Church where his son, Frederick I, was crowned in 1701, making Prussia a sovereign kingdom (see also p. 75).

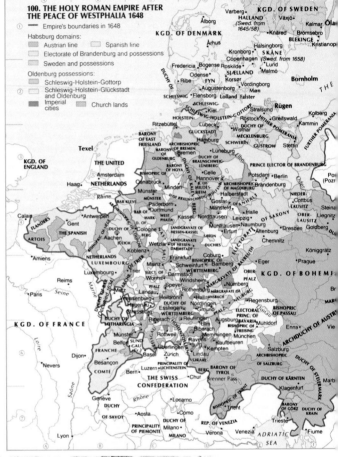

100. THE HOLY ROMAN EMPIRE AFTER THE PEACE OF WESTPHALIA 1648

— Empire's boundaries in 1648

Habsburg domains:
Austrian line   Spanish line
Electorate of Brandenburg and possessions
Sweden and possessions

Oldenburg possessions:
Schleswig-Holstein-Gottorp
Schleswig-Holstein-Glückstadt and Oldenburg
Imperial cities   Church lands

The Ottoman (Turkish) Empire 1360–1683

1360–89 Under Murat I the empire expands greatly in Asia Minor (map nr. 102).

1389–1402 Bajasid I Jildirim conquers large areas of the Balkans. Bulgaria is annexed in 1393.

1451–81 Mohammed (Mehmet) II captures Constantinople in 1453, making it the capital of the Osman Empire. Serbia, Bosnia, Albania and Greece become Turkish provinces.

1512–20 Selim I conquers the northern parts of Mesopotamia,

Tent encampment outside Vienna during Turkish siege in 1529.

Soldier with a hand mortar in Peter the Great's guard.

A host of crippled veterans were forced to beg for a living.

101. RUSSIAN EXPANSION 1300–1795

☐ Muscovy ca. 1300
☐ Muscovy 1462
☐ Lands conquered before 1689
■ Peter the Great's conquests 1689–1725
■ Conquered 1725–95
See also maps 120 and 126

The Russia of Peter the Great

Peter the Great, seen **above** in the battle against the Swedish king Charles XII at Poltava in 1709, was born in 1672, the son of Tzar Alexis by his second marriage. Following Alexis' death in 1676, there was a struggle for the throne among family members. Peter was proclaimed tzar when only ten years old, but he had no real power before 1689. During the nearly thirty years that Peter the Great governed the country, he opened Russia to western technology and customs. He had only been governing on his own one year when he undertook a lengthy period of travel in Western Europe, to learn ship building and other trades. And he brought western experts back with him to his backward country.

In the Great Northern War (1700–21) he took part on the side of Poland and Denmark against Sweden. The prize was Livonia, Estonia, Ingermanland and portions of Karelia (map nr. 101).

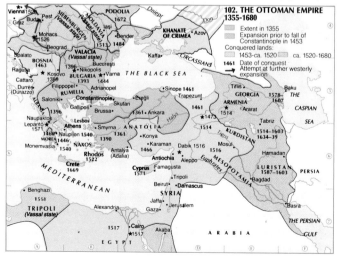

102. THE OTTOMAN EMPIRE 1355–1680

☐ Extent in 1355
☐ Expansion prior to fall of Constantinople in 1453
Conquered lands:
☐ 1453–ca. 1520   ☐ ca. 1520–1680
1461 Date of conquest
→ Attempt at further westerly expansion

W. Kurdistan, Syria and Egypt.
1520–66 Under Suleyman II the Ottoman Empire takes Beograd (1521), Rhodes (1522) and most of Hungary. In 1529, 300 000 Turks lay siege to Vienna, but the city is able to withstand the assault. Siebenbürgen, Moldavia, Georgia, Armenia and Tripoli become vassal states.
From 1567 Gradual disintegration of the empire. In the naval battle at Lepanto in the same year, however, Spanish and papal forces are victorious.
1683 Vienna is once again under siege.

English pillory, used at the beginning of the 1700s.

 *Condemned man is drawn and quartered in the square of a French town in 1757.*

 The Tower of London, 1641. Built in 1078 king's residence until 1509.

Above: *Child portrait of Maria Theresa, painted in 1726 by the Dane Andreas Møller. Maria Theresa was then only nine years old. Fourteen years later she became a reigning empress, and governed her Catholic kingdom with strength and wisdom for another forty years.*

Austria 1740–80

Following the death of Charles VI in 1740, there was a struggle over the lands of the Hapsburgs. Only after the War of Succession (1740–48), in which Prussia, France, Bavaria, Saxony and Spain were involved, was Maria Theresa (see above) finally able to draw the longest straw.

She abolished torture during interrogation, limited the power of the nobles and regulated the legal rights of farmers vis-à-vis the landed proprietors.

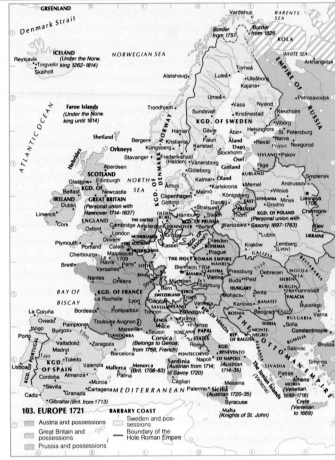

103. EUROPE 1721

- Austria and possessions
- Great Britain and possessions
- Prussia and possessions
- Sweden and possessions
- Boundary of the Hole Roman Empire

Left: *A meeting in the British lower house in London in 1710, i.e., three years after England and Scotland were united as Great Britain. From a contemporary painting by the Dutch painter Peter Tillemans. At this point the House of Commons had 558 members, and there were actually not enough seats for everyone many had to stand during legislative sessions. Fortunately, it was rare to find everyone present at once. Nearly half the representatives belonged to the landed nobility, The Gentry. Seated in the centre is The Speaker.*

Great Britain 1707–42

1707 England and Scotland are united as *The United Kingdom of Great Britain.* Following the Spanish War of Succession, Great Britain receives Gibraltar, Menorca, Newfoundland and Nova Scotia.

From 1714 Parliamentary system evolves. Industrial Revolution begins in the textile industry (map nr. 109, p. 77).

1742 Walpole forced to step down as PM due to failure in Austrian War of Succession.

Team of horses pulls wheeled plow, an 18th century invention.

Conscription of soldiers in the early 1700s.

Right: Road construction in France. From a painting by Joseph Vernet, now in the Louvre, Paris. The 18th century was a period of vast road building projects in France, and it was for the most part the peasants, obliged to provide the crown with day-labour, who shouldered the construction and maintenance of the royal highways. In addition, the state collected a toll from everyone who used roads and bridges. With the establishment of École des ponts et chaussées ('school of bridges and roads') technical standards rose enormously, and French highways were considered the best in Europe.

Poland 1764–95

1764 Catherine the Great of Russia secures the Polish throne for her former lover, Stanislav Poniatowsky.

1772 The first partition of Poland. Russia takes the territories east of the Daugava (Da) and the upper part of the Dnieper. Austria takes Galicia and Lodomeria. Prussia receives West-Prussia, with the exception of Danzig (Gdanzk) and Thorn (Torun) (map nr. 104).

1791 The Polish parliament – *sejm* – adopts a new constitution, making Poland a hereditary monarchy, and abolishing the right of free veto.

104. THE PARTITIONS OF POLAND 1772, 1793, 1795

Ceded to:	1772	1793	1795
Russia			
Austria			
Prussia			

▬▬▬ Poland before third partition
▬▬▬ Poland's boundary 1991

1792 In order to «reestablish order», Prussia and Russia undertake the second partition of Poland. Russia gets vast areas west of the Daugava and Dnieper. Prussia takes Danzig, Thorn, Poznan and South-Prussia (Kalisz).

1794 Tadeusz Kosciuszko leads a large rebellion, which is, however, put down by Russian and Prussian troops.

1795 The third partition of Poland. The country is obliterated as a nation, and is not declared an independent state again until 1917. Poland's final borders are not established until 1921.

The Russia of Catherine the Great 1762–96

Catherine the Great, shown here from a portrait by Alexander Roslin **below right**, was born Sophia Augusta, daughter of prince August of Anhalt-Zerbst. Born in Stettin in 1729, she was married at the age of sixteen to Grand Duke Karl Peter Ulrik of Gottorp. He was a nephew of the Russian empress Elizabeth, and succeeded her as Peter III when Elizabeth died in January of 1762.

In July of the same year, however, Catherine seized power in a coup, and only a few days later, Peter III died mysteriously.

In the course of her thirty-four year reign Catherine turned the country into a great-power. After a number of wars against the Ottoman Empire (Turkey) - 1768–74, 1783 and 1787–92 - Russia conquered the Khanate of the Crimean, and the coast of the Black Sea to the Dnestr (map nr. 120, p. 83). With the three partitions of Poland, Russia expanded her territory even more. Catherine wanted to be seen as a great friend of the philosophers of the Enlightenment, but seldom took their advice.

Basic steps in the minuet, most popular dance of the Rococo period.

Armchair from the late Rococo, ca. 1765.

Rococo woman of the upper classes, dressed in skirted gown, 1766.

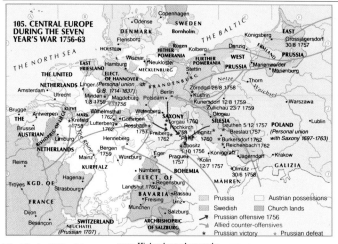

105. CENTRAL EUROPE DURING THE SEVEN YEAR'S WAR 1756-63

| | Prussia | | Austrian possessions |
| Swedish | | Church lands |
Prussian offensive 1756
Allied counter-offensives
★ Prussian victory ✶ Prussian defeat

The Musical Life of the 18th Century

The entire century was a period of flourishing musical activity. One of the greatest names was Wolfgang Amadeus Mozart (1756–91), who during his short life created music that still today is as alive, airy and life-affirming as when it was written over 200 years ago. Wolfgang Amadeus was the son of Austrian violinist and composer Leopold Mozart, and composed his first pieces of music at the age of six or

Below: Family concert in the palace of Renescure (18th century). Painting by an unknown artist. Musée des Beaux Arts, Lille.

seven. His best known large works are «Abduction from Seraillet» (1782), «Figaro's Weddings» (1786), and «Don Giovanni» (1787).
Other leading composers from this rich period were Johan Sebastian Bach (1685–1750) and Joseph Haydn (1732–1809).
In Italy both Antonio Vivaldi (1678–1741) and Domenico Cimarosa (1749–1801) were active. And in Great Britain, the German-born composer, Georg Friedrich Händel wrote one large oratorium after the other. In Sweden Carl Michael Bellman (1740–95) was writing and performing his «Fredman's Epistles» and «Fredman's Songs».

The Seven Years' War 1756–63

For some years already there had been confrontations between British and French troops in America, and between naval units at sea, when in 1756 Great Britain declared war on France. The French had at this time allied themselves with the Austrians, while Great Britain had a mutual defense pact with Frederick the Great of Prussia. Later, Russia, Spain, Saxony and Sweden were drawn into the conflict on the side of France in the war on the European continent, in which the largest battle was fought between Prussia and Austria-Russia.

Great Britain's greatest contribution to the clashes in Europe was its subsidizing of Prussia, who began the struggle with a surprise attack on Saxony in the autumn of 1756. One of the most famous battles of the war was fought at Rossbach in Saxony, on November 5, 1757, where Prussia's 21 600 man army was attacked by a combined force of 43 000 Germans, Austrians and Frenchmen. By a surprise manoeuvre the Prussians managed to surround the allies, who fled in panic into the biting cold of the winter's night. Frederick the

Below: Wolfgang Amadeus Mozart. From a contemporary painting by an unknown artist.

Man and wife harvesting together in 1675.

Harvest wagon. ca. 1750.

Wheelbarrow ca. 1750.

Left: From the battle of Prague, May 6, 1775. Contemporary copperplate engraving. The Austrians could not stop Frederick the Great's crack Prussian troops. They were driven back into the strongly fortified city of Prague. Still, Frederick paid a high price for victory. Nearly 18 000 of his men lost their lives, and the loss of four hundred officers – among them his able general, Kurt von Schwerin – was very nearly catastrophic «Schwerin was worth 10 000 men,» the king was heard to sigh.

Great's losses were 156 dead and 376 wounded. Losses to the allies numbered in the thousands.
Map nr. 105 shows Prussia's most important charges with red arrows, allied counterattacks in blue. Red stars indicate Prussian victories, blue stars those of the allies.
The most significant effects of the Seven Years' War, however, came about as a result of battles between

Below: Uniforms from Frederick the Great's army. From a small-scale plaque drawn by von Muhlen.

the British and the French that were fought out on battlefields far from Europe, and at the Peace of Paris in 1763, France was forced to cede her colonies in North America to Great Britain (map nr. 85, p. 62). The French possessions in India were also lost.

Frederick the Great

Frederick II, who was later given the nickname «The Great», was in his youth – to his father's dismay – greatly influenced by French Enlightenment philosophy. His command of French surpassed that of his own mother-tongue, and he corresponded regularly with the greatest authors of the Enlightenment in France, François Voltaire, and others. He wrote poetry in French, and composed around 120 pieces for the flute.
When Frederick II took the reins of government in Prussia in 1740, he further bolstered the country's military so that Prussia became the fourth largest military power on the continent.

Above: Frederick the Great of Prussia, the king who divided his energies between war and philosophy. Painting by J.G. Glume.

Brandenburg-Prussia
1415–1795

1415 Brandenburg comes under the Royal House of Hohenzollern.
1525 End of the rule of the Teutonic Knights, and East-Prussia becomes a duchy under Polish hegemony.
1539 The Lutheran religion is introduced.
1618 Prussia is united with Brandenburg (map nr. 101, p. 70).
1640–88 Under «the Grand Electoral Prince» – Frederick Wilhelm – both Further Pomerania and West-Prussia are placed under Brandenburg, and East-Prussia no longer Polish fief.
1701 Electoral Prince Frederick III of Brandenburg is proclaimed king of Prussia as Frederick I.
1720 Peace of Stockholm. Sweden cedes Pomerania to Prussia.
1742 Under Frederick II (the Great) Austria cedes Silesia.
1756–63 During the Seven Years' War, Prussia is on her way to becoming a great-power.
1772–95 Following the three partitions of Poland (map nr. 104, p. 73), Prussia receives large territories.

106. EXPANSION OF BRANDENBURG-PRUSSIA 1415–1797

Brandenburg 1415
Added 1415-1535
Added 1608-19
Added 1640-88
Added 1688-1740
Added 1740-86
Added 1786-97
1648 Year of annexation

★ Battlefield
— Boundary of the Holy Roman Empire

Woman from nothern
Ireland at spinning
wheel in the 1780s.

The Glyptotek in
Munich, and example
of Neo-Classicism.

Tenement buildi
in Paris in the mi
1700s.

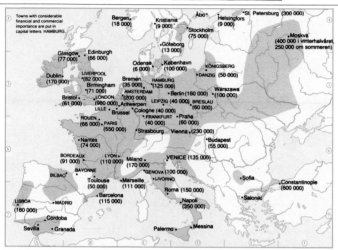

Towns with considerable
financial and commercial
importance are put in
capital letters: HAMBURG.

Bergen.
(18 000)

Kristiania
(9 000)

Åbo

Helsingfors
(9 000)

St. Petersburg (300 000)

Stockholm
(75 000)

Moskva
(400 000 i vinterhalvåret,
250 000 om sommeren)

Göteborg
(13 000)

Glasgow
(77 000)

Edinburgh
(66 000)

Odense
(6 000)

København
(100 000)

KÖNIGSBERG

Dublin.
(170 000)

LIVERPOOL
(82 000)

Bremen
(35 000)

HAMBURG
(125 000)

DANZIG (50 000)

Birmingham
(71 000)

AMSTERDAM
(200 000)

Berlin (160 000)

Warszawa
(100 000)

Bristol
(61 000)

LONDON
(980 000)

LEIPZIG (40 000)

BRESLAU
(60 000)

LILLE.

Antwerpen.

Cologne (40 000)

Brussel

FRANKFURT
(40 000)

Praha
(60 000)

ROUEN
(66 000)

PARIS
(550 000)

Strasbourg

Vienna (230 000)

Nantes
(74 000)

Budapest
(55 000)

BORDEAUX
(91 000)

LYON
(110 000)

Milano
(170 000)

VENICE (135 000)

BAYONNE

GENOVA (100 000)

BILBAO

Toulouse
(50 000)

Marseille
(111 000)

LIVORNO

Sofia

Constantinople
(600 000)

Barcelona
(115 000)

Roma (150 000)

Saloniki

LISBOA
(180 000)

MADRID

Napoli
(350 000)

Cordoba

Sevilla

Granada

Palermo.

Messina

107. POPULATION DENSITY IN EUROPE CA. 1800

Inhabitants per square kilometre

More than 40

20-40

Less than 20

Populations of certain towns ca. 1800 in brackets

The Population of Europe around 1800

The population of Europe had reached over 40 inhab. per sq. km in England, Belgium, The Netherlands, and in parts of Ireland, France and Italy (map left). Today, the population density in, f.ex., England and The Netherlands is more than 250 per sq. km, while in Great Britain generally, it is approximately 230 per sq. km. In the 1800s the pop. rose from 9 to 33 mil. in England, but only from 27 to 33 mil. in France. Russia had the greatest increase, from 39 to 110 mil. in the same period (excluding Poland and Finland).

London ca. 1800

The most densely populated districts in England around 1800 were the industrial areas around Liverpool, Manchester and Leeds (map nr. 109). But London, with almost a million inhabitants, was by far the country's largest city. **The picture below**, by Pugin and Rowlandson, shows a fire in London in 1808. The city's first fire brigade was sponsored by insurance companies, but in 1833 the London Fire Brigade was organized. St. Paul's Cathedral is seen in the background.

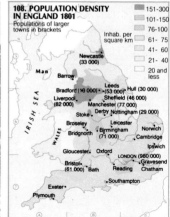

108. POPULATION DENSITY IN ENGLAND 1801

Populations of larger towns in brackets

Inhab. per square km

151-300

101-150

76-100

61- 75

41- 60

21- 40

20 and less

Newcastle
(33 000)

Man

Barrow

Bradford (16 000)

Leeds
(53 000)

Hull (30 000)

Liverpool
(82 000)

Manchester (77 000)

Sheffield (46 000)

Stoke

Derby

Nottingham (29 000)

Broseley

Leicester

Bridgnorth

Birmingham
(71 000)

Norwich

Cambridge

Gloucester

Oxford

Ipswich

LONDON (980 000)

Bristol
(61 000)

Bath

Reading

Gravesend

Chatham

Exeter

Southampton

Plymouth

IRISH SEA

WALES

The street-lamp is lit in a London street in 1800.

A handcart on the waterfront in Philadelphia in 1800.

A European enjoys a clay pipe of Virginia tobacco, ca. 1800.

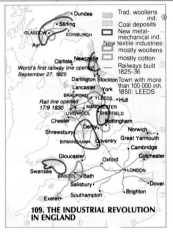

Trad. woollens ind.
Coal deposits
New metal-mechanical ind.
New textile industries:
   mostly woollens
   mostly cotton
Railways built 1825-36
Town with more than 100 000 inh. 1850: LEEDS

World's first railway line opened September 27, 1825

Rail line opened 17/9 1830

109. THE INDUSTRIAL REVOLUTION IN ENGLAND

Above: On many of the rivers and canals in England, barges were drawn by horses, following a path along the bank. From a painting by the famous landscape painter John Constables from around 1800.

Left: The two largest cities on the European continent around 1800 were Constantinople, with around 600 000 inhabitants, and Paris, with approximately 550 000. This detail from a plan of Paris by Minister of Finance Turgot from the last half of the 18th century shows central portions of Paris, with the Tuileries (tile works) and Pont Royal in the foreground.

Roads and Transport

During the fifteen years between 1765 and 1780, something of a transportation revolution occurred in France. As shown in map nr.

110, a journey from Brussels to Toulouse took nineteen days in 1765, as did a trip from Strasbourg to Rennes. Following a period of highway improvements, and the introduction of regularly scheduled

diligence routes (see Charles Rossiter's painting, below left), the trip from Brussels to Toulouse was reduced by 1780 to only eleven days, with stops. From Rennes to Strasbourg it now only took 8 days.

Above: George Stephenson's steam locomotive – invented 1813/14 – pulling a row of coal cars. This invention was crucial to the transport needs of the new industrial areas in northern England.

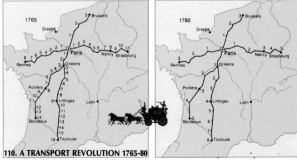

110. A TRANSPORT REVOLUTION 1765-80

French gallantry among the upper classes at the end of the 1700s.

Women too took part in the revolution in 1789.

Rebellious soldiers with various weaponry in Paris 1789.

IRELAND
Nationalist uprising against the British, led by Wolfe Tone

THE UNITED NETHERLANDS (1784)
THE AUSTRIAN NETHERLANDS (1787)
Paris

FRANCE
The French Revolution
Storming of the Bastille July 14, 1789.

Madrid *SPAIN
A popular revolt against French rule leads in 1808 to a broadly based guerilla war.

POLAND
Rebellion against the Russians in 1794, led by general Tadeusz
Warszawa ★Kosciuszko
Maciejowice
★ Raclawice

HUNGARY
Budapest • Ignac Martinovics heads a Jacobin-Republican conspiracy against the monarchy in 1794.

TYROL
Uprising in 1808/09, led by Andreas Hofer after Bavaria had occupied Tyrol.
Beograd

SERBIA
Peasant rebellion against the Turks in 1804, led by Karajordje (Black George)

SOUTHEAST RUSSIA
Peasant uprising 1773-75, led by the Cossack J.I. Pugachev, who disguised himself as the dead Tzar, Peter III.
Saratov

Tsaritsyn

111. REVOLUTION AND FREEDOM IN EUROPE CA. 1800

Above: The actor Chinard in the sans-coulottes costume. The petty bourgeoisie, the radical force of the Revolution, wore long trousers, in contrast to those worn by the bourgeoisie proper. They were sans coulotte, i.e., without knee breeches. Painting by Louis-Léopold Boilly.

European Revolutions

1773–74 Yemelyan I. Pugachev leads a peasant uprising in southern Russia (see also p. 83).

1789 The French Revolution (see next column).

1794 Ignac Martinovich heads a conspiracy against the monarchy in Hungary.

1794 General Tadeusz Kosciuszko leads a rebellion against the Russians in Poland (see p. 73).

1796–98 The Irish nationalist, Wolfe Tone directs an unsuccessful war of independence.

1804 In Serbia, prince George Petrovich (Black George) leads uprising against the Turks.

1808 An uprising in Spain leads to a protracted guerilla war against the army of Napoleon.

1808–09 The Tyrolean freedom fighter Andreas Hofer drives the French from Tyrol.

Below: In 1788 censorship was abolished in France. The presses were spewing out newspapers and pamphlets.

The French Revolution 1789–95

1789

May 5 The Estates General meet for the first time since 1614. Louis XVI has been forced to convene the assembly because of an economic crisis.

June 17 The Third Estate forms a constituent national assembly (the **Constituante**), and three days later they pledge to separate until France has a new constitution (the Tennis Court oath).

July 14 Storming of the Bastille (see illust. on the following page).

July 20 – August 5 Peasant uprisings throughout most of France. Rumours that bands of robbers are plundering the country cause the uprisings to be referred to as «The

Great Terror» (map nr. 113, p. 79). The clergy and the nobility relinquish their privileges.

October 5 Because of the lack of food in Paris, a procession of market woman march to the palace of Versailles, chanting «Bread for Paris!» (illust. following page).

1791 Fall of the monarchy. Louis XVI and Marie Antoinette flee, but are stopped.

1793–95 The royal couple is executed (January 21 and October 17, 1793). Uprising of royalists and federalists (map nr. 114, p. 79). The «Committee of Public Safety» organizes mass executions. Nine states form a coalition and declare war. France wins key victories. Among the foremost military leaders is Napoleon Bonaparte.

112. PARIS AT THE OUTBREAK OF THE FRENCH REVOLUTION 1789

A woman and man of the Third Estate toast the Revolution in 1789.

Representatives of Paris' light-hearted leisure class in the 1790s.

Louis XVI is executed by guillotine in Paris, on January 21, 1793.

Right: On October 5, 1789 between five and six thousand women walked the 30 km from Paris to Versailles to demand bread.

Below: Storming of the Bastille, July 14, 1789. From a contemporary colour lithograph. When the commandant saw the mass of people approaching, he gave the order to open fire. Then, following a short but hectic battle, he raised the white flag, and the people stormed in. The prisoners, seven in all, most of them criminals, were freed, and the crowd marched in triumph through the streets of Paris, carrying the head of the commandant ⟨on a⟩ pike. July 14 became Fran⟨ce's⟩ of national celebration (⟨Bastille⟩ Day).

113. FRANCE UNDER THE TERROR
- ➾ Where the uproar started
- ☐ Unaffected areas
- ▨ Peasant revolt July 1789

Brussel · Cologne · ARTOIS · Maas · Koblenz · Amiens · PICARDIE · Sedan · Mosel · Worms · Le Havre · Estrée · Jersey · Rouen · Valmy · BRETAGNE · NORMANDIE · PARIS · CHAMPAGNE · Versailles · Romilly · Rennes · Laferté · Troyes · ANJOU · Orléans · Saint-Florentin · Nantes · POITOU · BERRY · NIVER-NAIS · FRANCHE-COMTE · Bern · La Rochelle · MARCHE · Louhans · AUVERGNE · Lyon · Torino · Ruffac · LIMOUSIN · Rhône · DAUPHINE · Bordeaux · GIRONDE · Garonne · Dordogne · LANGUEDOC · Avignon · PROVENCE · Bayonne · Pau · Toulouse · Marseille · Nizza (Nice) · Carcassonne · Toulon · MEDITERRANEAN

114. REVOLT IN THE PROVINCES
- ▨ Royalist, Summer 1793
- ☐ Federalist, Summer 1793
- — France's boundaries 1792
- ▨ Conquered 1794-95

Haag · Nijmegen · Dunkerque · Antwerpen · Neerwinden · Boulogne · Fleurus · Koblenz · THE ENGLISH CHANNEL · Valenciennes · BELGIUM · Arras · Le Quesnoy · Mainz · Amiens · Varennes · Brest · Rouen · Saint-Denis · NORMANDY · PARIS · Valmy · BRETAGNE · Versailles · Seine · Quiberon · Rennes · Basel · Savenay · Loire · Orléans · Châtillon · Montbéliard · Nantes · Bourges · Dijon · FRANCHE-COMTE · VENDÉE · Nevers · La Rochelle · Genève · Lyon · SAVOIA · Bordeaux · Mende · Valence · KGD. OF SARDINIA · Garonne · Avignon · PROVENCE · Bayonne · Dax · Nîmes · Nice · Toulouse · Montpellier · Marseille · Foix · Toulon · SPAIN · ROUSSILLON · THE MEDITERRANEAN · Perpignan

Emperor Napoleon in a characteristic stance.

French canon from the 1760s. Used during the Napoleonic Wars.

Empress Joséphine, whom Napoleon divorced in December 1809.

WÜRTTEM-BERG

KGD. OF BAVARIA (From 1808)
· Konstanz

·Zürich

FD. NEUENBURG (NEUCHÂTEL) (Prussian)

HELVETIAN REPUBLIC (1798–1803)

·Geneve ·Martigny

Great St. Bernhard

·Vienna

STEIERMARK

Buda ·· Pest

AUSTRIA

SALZBURG
KÄRNTEN

Campoformio
Peace Treaty Oct. 18 1797

SAVOY. (Occupied by Fr. 1792)

PIEMONTE (To Fr. from 1796)
·Sardinia 1796

SAXONY OF TYROL
To Kgd. of Italy 1810

·Udine

KRAIN

·Trieste

Drava

·Zürich

LOMBARDIA
·Milano
Novare ·Lodi
Marengo 14/6 1800
Mondovi 22. 4 1796
Millesimo 13/4 1796
Nizza (Nice) ·Monaco

14/1 1796
Arcole 17.11 1796
10/5 1796

Rivoli
Bassano 8.9 1796

·Torino
·Genova

Piacenza
·Parma

Mantova (Besieged June 1796 – Feb. 1797)
Guastalla
THE CISALPINE
·Modena ·Bologna
REPUBLIC

·Venezia

ISTRIA

·Pola

REPUBLIC OF VENEZIA

CROATIA

Sava

BOSNIA

SERBIA

DUCHY OF PARMA
·Ferrara

THE LIGURIAN REP. (1797–1805)

·Rimini

·Zara

DALMATIA

THE OTTOMAN EMPIRE

HERCE-GOVINA

·Sarajevo

LUCCA
·Firenze
SAN MARINO
·Ancona

TOSCANY
Napoleon arrives with his army March 26, 1796

·Pisa
·Livorno

KGD.
OF ETRURIA
(1801–08)
Piombino ·Elba

·Siena
PAPAL
·Perugia
·Assisi
STATES

·Orvieto ·Spoleto

REP. OF RAGUSA

MONTE-NEGRO

THE ADRIATIC SEA

Corsica (French from 1768)

Orbetello
(THE ROMAN REPUBLIC 1798–99)

·Ajaccio

·Roma

PONTECORVO

KGD.
·Barletta

Gaeta

B BENEVENTO

·Bari

·Capua

·Salerno

·Napoli

OF NAPLES

(THE PARTHENOPEAN REP. 1799)

·Brindisi

KGD. OF SARDINIA

TYRRHENIAN SEA

·Otranto

·Cagliari

Liparian Islands

Palermo

Messina · Reggio

115. NAPOLEON'S ITALIAN CAMPAIGNS 1796–1800 AND FRENCH DOMINATION IN ITALY CA. 1800

French territory:

France 1792	Luccan Rep. 1799–1805
Piedmont, taken from Sardinia 1798	Kgd. of Naples. Parthenopean Rep. 1799
Duchy of Parma, annexed 1803	Kingdom of Etruria 1801–08

Under French control:

Cisalpine 1797–02, Italian 1802–05	Duchy of Piombino 1801
Ligurian 1797–05	Kingdom of Italy 1808
Papal States, Roman Rep. 1798–99	

Napoleon's campaigns:
1796–97 — 1800
Venetian Republic, Austrian 1797–1805

Marsala

KGD. OF SICILY

·Catania

·Syracuse (Syrakus)

Malta (French 1798–1800)

Napoleon 1796–1800

1796–97 In March 1796 the twenty-six-year old Napoleon Bonaparte marries Joséphine de Beauharnais, six years his senior, with whom he is passionately in love. The fact that she had good political connections did nothing to lessen his passion.

A few weeks later he is placed in command of the French army, which was set to attack the Austrians in northern Italy. He departs from Nice and crosses the border into Piemonte, and in April and May of 1796 he defeats the Austrians in a series of battles: Millesimo, Dego, Mondovi, Lodi (map nr. 115). After victories at Bassano in September, Arcole in November, and the fall of Mantua in February 1797 peace is negotiated in Campoformio. Lombardy becomes The Cisalpine Republic.

1798 Having given up plans to break Great Britain with an attack across the Channel, Napoleon moves against British Egypt in May. On July 21 he defeats the Mameluks and captures Cairo, but when Admiral Nelson gives the British a victory in a naval battle at Aboukir (illust. p. 81), Napoleon and his army are trapped in Egypt. He pushes on toward Syria to defend against a Turkish offensive, but is forced to return to Egypt.

1799 N. leaves Egypt, arriving in Paris at the end of October. Carries out a coup d'etat, and is made First Consul with very nearly dictatorial powers.

1800 French forces are trapped in Genoa, and Napoleon decides to go to their rescue. In May he leads an army of 60 000 men over the Alps (illust. above), and following his victory at Marengo on June 14, he can boast «Italy's fate is sealed.» (cont. p. 82)

Above: The Corsican, Napoleon Bonaparte, who had become a lieutenant in the artillery in 1785 at the age of 16, is, in this portrait from 1796, a general, leading the French forces to victory in a battle against the Austrians at Arcole northwest of Venice on November 17, 1796. It was here that he inspired his troops by charging ahead of them into the fray with a raised banner. From a painting by Jean Gros.

Empress Joséphines bed in the palace of Malmaison.

Mens' and ladies' dress in Napoleon's time.

Fontainebleau where Napoleon said farwell to his soldiers in 1818.

116. EUROPE AFTER THE PEACE OF LUNÉVILLE 1801

France and the four "republics"
Habsburg domains
Prussia and possessions
Boundary of Holy Roman Empire

Above: Napoleon's army crossing the Alps, May 15–20, 1800. The troops have reached San Bernadino, where they dismantle their canons and pack them in hollowed out tree trunks for transport through the pass. From a contemporary painting by Charles Thévenin.

The Battle of Austerlitz 1805

On the anniversary of his coronation as emperor, December 2, 1805, Napoleon defeated Tzar Alexander I's forces, and what was left of the Austrian army, at Austerlitz. This was the crushing blow for Austria.

The Peace of Lunéville 1801

Following Napoleon's victory at Marengo (see p. 80) Franz II of Austria sued for peace. It was concluded in Lunéville February 9, 1801, and stated among other things that the Rhine would form France's eastern border, from Switzerland to The Netherlands. Austria was also forced to recognize the four sister republics, the Ligurian, the Cisalpine, the Helvetian, and the Batavian (map nr. 116). France had thus consolidated her power in Europe, as had First Consul Napoleon in France.

Right: England's famous admiral, Horatio Nelson, Napoleon's nemesis at Aboukir in 1798, and at Trafalgar in 1805.

Napoleon in Spain and Portugal 1808–14

In 1807 Napoleon occupied Portugal. In May 1808, he deposed the Spanish king and placed his own brother, Joseph, on the Spanish throne. Only a short time later, however, all of Spain was in revolt. The Duke of Wellington came to the rescue with 8 000 British soldiers. (map nr. 118).

117. BATTLE OF AUSTERLITZ 1805

Marshal Lanne's infantry
Murat's cavalry divisions
Napoleon's garde
Marshal Bernadotte
Marshal Soult
Marshal Davout's infantry divisions

General Bagration's Russian forces
Prince Lichtenstein's German cavalry
The Russian Tzar's Imperial Garde
AUSTERLITZ
General Vandamme
General St Hilaire
General Kollowrat's Russian forces
General Buxhöveden's Russian forces

■ French forces
□ Russians and Austrians
Marsh

0 1 2 3 4 5 km

118. THE FRENCH CAMPAIGNS IN IBERIA 1808-14

→ Napoleon's thrust
→ British counter-offensives
★ Battlefield

William Pitt and Napoleon compete for the largest slice of the globe.

Napoleon's simple throne is in the Neo-Classical style – 'Empire'.

Napoleon returning to Paris after defeat in Russia in 1812.

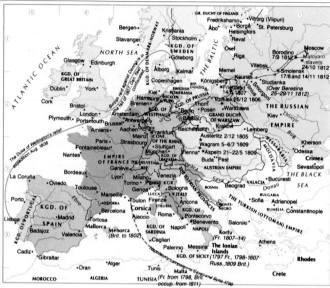

119. EUROPE IN 1812

- French territory
- Ruled by members of Napoleon's family
- States under French control
- Allies of France against Russia in 1812
- Neutral states
- The Rhine Confederation
- → Napoleon's Moscow campaign, May-December 1812

tion of the Rhine, with Napoleon as protector (map nr. 119). Three months later Napoleon marches on Prussia, and on October 27 he sets up his headquarters in Frederick the Great's Potsdam. On Nov. 21 the mainland blockade of Great Britain is initiated. After the Peace of Tilsit in July 1807, France is Europe's most powerful state.

1808 The French occupy Portugal and Spain (see p. 81).

1812 Napoleon marches toward Moscow (map nr. 119) and occupies the city, but without any final capitulation. During the French withdrawal, great numbers of Napoleon's troops perish.

1814 N. abdicates on April 7, and leaves for Elba, left to him by the charitable victors (below).

Napoleonic France 1801–14

1801 Peace is concluded with Austria at Lunéville on February 9 (see p. 81). On July 15 Napoleon signs a concordat with Pope Pious VII, stating that Catholic mass would once again be allowed, following a ten-year ban.

1802 The Peace of Amiens, March 27, puts an end to the war with Great Britain. The results of a plebescite make Napoleon First Consul for life.

1803 On May 16 Great Britain again declares war on France.

1804 Napoleon is elected emperor in May, and on December 2 the coronation takes place in Notre Dame in Paris. Spain joins France in the war against Great Britain.

1805 Great Britain, Russia, Austria, Sweden and the Kingdom of Napoli (Naples) form the Third Coalition against France. The goal of the alliance is to force France back behind its former borders. On October 20 the French defeat the Austrians at Ulm, but the very same day Admiral Nelson defeats the French fleet at Trafalgar. On November 13 Napoleon enters Vienna. The battle of Austerlitz takes place on December 1–2 (see p. 81).

1806–07 On July 16, 1806 sixteen German states form the Confedera-

From a Russian village at the beginning of the 1800s.

Russian coach from around 1800.

Russian serfs get a taste of the whip around 1800.

Above: *Alexander I, Russian Tzar 1801–25. From James Walker's portrait. Alexander I was at first a liberal monarch, but became increasingly conservative – not least after Napoleon's invasion in 1812.*
Left: *Napoleon says farewell to his officers and men in the palace square outside Fontainebleau before his departure for Elba, April 20, 1814. From a painting by Horace Vernet, in Fontainebleau.*
Below: *The Don Cossack Yemelyan Pugachev, masquerading as the murdered Tzar Peter III, led an uprising in southern Russia in 1773–74 (map nr. 120). Here Pugachev is being transported in a cage to Moscow, where he was eventually executed.*

Russia 1796–1812

1796 Tzar Paul I – a bitter, introverted man, totally unsuited to rule his vast empire – succeeds his mother, Catherine the Great (see p. 73). In the nearly four years of his reign, Paul I wavers back and forth from one ally to the other.
1801 In March Tzar Paul is deposed in a coup d'etat and killed. His son, Alexander I Paulovitch (illust. left), who has consented in the coup, takes over.
1805 Russia joins the Third Coalition against Napoleon, but suffers defeat at Austerlitz (see p. 81).
1807–09 Following yet another defeat at Friedland in June 1807, Alexander I makes peace with Napoleon in Tilsit. Russia must promise to join the mainland blockade, and cede territories to Napoleon's dependency, the Grand Duchy of Warsaw (map nr. 119, p. 82). On the other hand, the Russians are given a free hand in Finland, which is conquered in 1808–09 and made a Grand Duchy.
1812 Bessarabia is taken from the Ottoman Empire (map nr. 105). Napoleon attacks Russia after Tzar

Alexander has resumed trade with a deserted Great Britain. The French reach Moscow on September 14, and Napoleon sets up headquarters in the Kremlin. The following day the Muscovites set fire to the city, where 80 percent of Moscow's 9 000 buildings are made of wood. Napoleon offers the Tzar a ceasefire, but the offer is rejected. On October 18 Napoleon orders his army to pull out. Of the 610 000 men who began the campaign in June, only 85 000 make it home.

120. THE RUSSIAN EMPIRE 1762-1812

Russia in 1762
Conquests 1762-96
Russia in 1812
Affected by Pugachovs revolt 1773-74

British visiting an occupied Paris in the summer of 1814. Caricature.

Napoleon at the head of his army in 1814.

Louis XVIII has returned to Paris after fleeing in 18

Napoleon Will Not Surrender

He had been allowed to retain the title of emperor, and a force of 1 200 men, which he still drilled regularly. But Napoleon was suffering from boredom on Elba.

At the end of February 1815 – more than five months after the Congress of Vienna had convened (see p. 85) – an opportunity for further dramatics presented itself. Leading his men, he managed to reach Antibes undetected. March 10 he reached Lyon. Three days later the Congress declared him an outlaw, and a large number of countries vowed that they would not lay down their arms until he had been rendered harmless.

On March 19 Louis XVIII fled Paris (illust. p. 85), and the next day – cheered by his veterans – Napoleon once more occupied the Tuileries. On June 1 he was sworn in, promising a new, more liberal constitution.

Seventeen days later he was defeated at Waterloo (see p. 85). Napoleon was transported to Great Britain on July 15, 1815 (**illus. above**), and then on to St. Helena, where he ended his days on May 5, 1821.

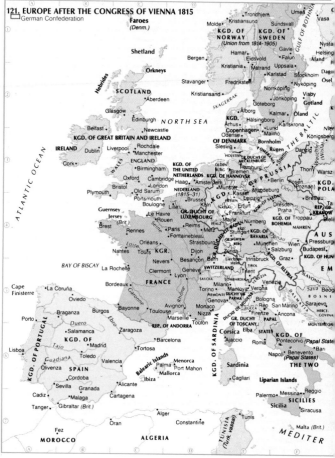

121. EUROPE AFTER THE CONGRESS OF VIENNA 1815
German Confederation

Left: Some of the delegates to the Congress of Vienna, 1814–15. In the left foreground sits Prussia's negotiator, Karl August von Hardenberg. The gentleman to his right, in front of the chair, is Austria's chief negotiator, Prince Metternich. Russia's Carl Vasilyevitch von Nesselrode is seen standing just to the right of Metternich. The man with his legs crossed – in the centre of the picture – is the British foreign minister, Lord Castlereagh. Farthest right – with his arm on the table – is France's foreign minister, Talleyrand.

Delegates to Congress of Vienna «dance and dance. but get nowhere.»

280 kg of silver were used to make the cradle for Napoleon's son.

Parisian girls of the «light guard» getting ready for the evening.

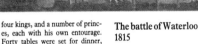

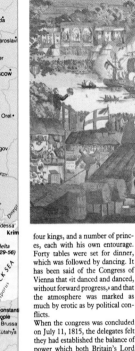

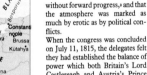

The Congress of Vienna 1814–15

From the French Revolution in 1789 until the Congress of Vienna convened September 16, 1814, the political situation in Europe had changed drastically. In Germany and Italy over 250 city-states and principalities had been dissolved and replaced by new ones. It was these conditions that the Congress of Vienna was supposed to put in order.

The delegates were all men, but in the evenings the ladies took the spotlight. Among emperor Franz's guests at Hofburg were **one** Tzar,

four kings, and a number of princes, each with his own entourage. Forty tables were set for dinner, which was followed by dancing. It has been said of the Congress of Vienna that «it danced and danced, without forward progress,» and that the atmosphere was marked as much by erotic as by political conflicts.

When the congress was concluded on July 11, 1815, the delegates felt they had established the balance of power which both Britain's Lord Castlereagh and Austria's Prince Metternich had been ardently promoting. France was forced to withdraw back behind her old borders, and Belgium was thrown together with Holland in The United Netherlands (map nr. 121). Russia received three fourths of the Grand Duchy of Warsaw, the so-called «Congress Poland», with Tzar Alexander as king. Prussia was also expanded, although divided in two along religious lines. The many German states were joined in a loose federation – The German Confederation (Deutscher Bund).

Right: On March 19, 1815 Louis XVIII, now sixty years old, is again forced to flee. This time he is accompanied by his and Louis XVI's younger brother, who would become Charles X (farthest left). The royal pair are making a hasty retreat with the crown jewels. Anonymous French caricature from 1815.

The battle of Waterloo 1815

On March 6, 1815 the members of the Congress of Vienna received word of Napoleon's return from Elba (see p. 84). They reacted immediately. In less than an hour a coalition had resolved to carry out another military campaign against him.

The final clash took place at Waterloo, a small village in Belgium, on June 18, 1815. In spite of the emperor's gifts as a tactician, Napoleon's 125 000 men were forced to surrender to Wellington and Blücher's 220 000 man force.

Above: Following Napoleon's abdication on April 7, 1814 (see p. 83), an interim peace treaty was signed in Paris on May 30, 1814. One result of this peace was that the British forces could immediately return home. The peace was celebrated with a huge fireworks display in Hyde Park in London, August 1, 1814. This hand-painted copperplate commemorating the event was issued only a couple of weeks after the festivities. A month later the Congress of Vienna was convened to work out the details of the peace treaty of May 30.

A «petroleuse» – arsonist – during the uprising in Paris in the spring of 1871.

A crinoline clad lady putting a letter in a Paris post box, from 1850.

Three pupils before their headmaster. Silhouette from the 1800s.

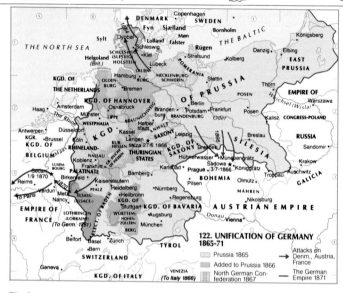

122. UNIFICATION OF GERMANY 1865-71

Prussia 1865

Added to Prussia 1866

North German Confederation 1867

→ Attacks on Denm., Austria, France

The German Empire 1871

tion (map nr. 121, p. 84).

1864 Prussia and Austria attack Denmark in January, taking Schleswig and Holstein (map nr. 123).

1866–67 Disagreement over the administration of Schleswig-Holstein leads to war between Prussia and Austria in the summer of 1866. Following Prussia's victory, the German Confederation is dissolved, and in 1867 the North German Confederation is formed, under the leadership of Otto von Bismarck (map nr. 122).

1870 Bismarck arms the country (see illus. bottom) and provokes the French under Napoleon III, until on July 19 France declares war on Prussia. Following a humiliating defeat at Sedan, Napoleon III surrenders to Bismarck. In France the imperial dynasty is abolished. The Republic is reintroduced. The war continues.

1871 On January 28 the French capitulate. By this time king William of Prussia has already become German emperor, and Bismarck chancellor (illus. below). At peace negotiations in May, France has to cede Alsace-Lorraine (Elsass-Lothringen).

The German Empire 1859–71

1859 In a war with France and Sardinia (map nr. 124) Austria loses Lombardy and Parma. At the same time, Prussia is gaining influence in the German Confedera-

123. SCHLESWIG-HOLSTEIN

Danish areas in Schleswig 1864

Areas ceded to Denmark, Treaty of Vienna Oct. 1864

→ Austro/Prussian attack- in all 51000 troops

Schleswig's northern boundary:
- —— 1864 – – – after 1864
- —— Boundary of German Confederation 1815-66
- —— Denm.'s south boundary 1920

Above: William I is proclaimed emperor of Germany at Versailles on January 18, 1871. The picture's central figure is Prussia's formidable prime minister, Otto von Bismarck. Beside him stands Field Marshal General Helmuth von Moltke. From a painting by Anton von Werner from 1876.

Left: Prussian soldiers test new canons from the Krupp Munitions Works at a firing range outside Berlin. At the right, Otto von Bismarck is seen on horseback. The man in the top hat is presumably Alfred Krupp. Woodcut in «L'Univers Illustré» from January 2, 1869.

Garibaldi helps Victor Emanuel II put on the «Italian boot».

Sofa for two persons, a so-called «vis-à-vis» from the 1860s.

A French student and his «grisette» – a self-employed woman.

The Apennines (Italian) Peninsula 1815–61: Italian Unification

1815 Following the Congress of Vienna, French hegemony is transferred to Austria, led by Metternich. The kingdom of The Two Sicilies comes into being (map nr. 124).

1830 The despotic Ferdinand II, a close relative of the Austrian emperor, becomes king.

1831 Giuseppe Mazzini founds the revolutionary liberation organization *Giovine Italia*, Young Italy, whose main political goal is a united Italian republic.

1848–49 Uprisings in Naples and Palermo lead to a new and more liberal constitution. Sardinia, the Papal States and Toscany follow suit, and after the revolution in Vienna in 1848, Milano and Venice break with Austria. Only a year later, however, the Italian states are again brought to heel.

1859 France and the Kingdom of Sardinia form an alliance against Austria. After the Battle of Solferino (see below) Lombardy and Parma fall to Sardinia.

1860 Giuseppe Garibaldi (see illust. at right) and his armed volunteers, the «Red Shirts», who carried out the «Campaign of the 1000», go ashore in Sicily in May (map nr. 124), and in September they take Naples.

1861 Garibaldi's men advance northward, as Sardinia's forces push southward. Central Italy falls, and on March 17 the Kingdom of Italy is proclaimed, with Sardinia's Victor Emanuel II as its first king.

Above: Following the French February Revolution in 1848, Napoleon I's nephew, Louis Napoléon Bonaparte was elected president. In 1852 he became Emperor Napoleon III, shown here as a prisoner in the castle of Wilhelmshöhe following his defeat at Sedan in 1870. Bismarck is the consoling victor.

Below: On June 24, 1859 Napoleon III led combined French and Sardinian forces against the Austrians at Solferino. With a total of 30 000 killed and wounded in one day, the battle was one of the bloodiest of the 19th century, and gave rise to the establishment of the International Red Cross.

Above: Freedom fighter Giuseppe Garibaldi. From an unsigned Italian painting. Garibaldi struggled all his life for freedom and justice, as in South America, where he led the defense of Montevideo against an Argentinean attack in 1843. During the unification of Italy in 1860-61, he was Victor Emanuel's most important supporter.

124. UNIFICATION OF ITALY 1859-61
- Sardinia 1859
- Austrian 1859
- Sardinia in May 1860
- Added to Sardinia/Autumn 1860
- Garibaldi's expedition 1860

A newspaper reader in Paris around 1830.

Women of a harem in a Turkish bath. From at painting by Ingres from 1862.

A taste for things Oriental made an impact on men's fashion too.

125. GREEK WAR OF INDE-PENDENCE 1821-29

Turkish in 1821

Ionian Islands, republic under British protection 1815-63

Kingdom of Greece, recogni-zed by great powers 3/2 1830

Below: Euge Delacroix' famous revolutionary painting «July 28. Liberty leads the people» from 1830 – now in the Louvre – symbolizes one of the 19th century's characteristic features: The people's desire for national liberation. This was certainly the case in France, Italy and Greece.

Above: George Noel Gordon, from 1798: Lord Byron (1788–1824). From a painting by Thomas Phillios. Byron's poetry, and his ardent belief in political freedom, was a source of inspiration for poets and liberation movements alike, not least in Greece. Fired by his enthusiasm for heroic deeds, the romantic Lord Byron travelled to Greece to take part in the struggle against the Turks. A few months after his arrival, however, he came down with rheumatic fever, and died on April 24, 1824. The unusually handsome poet became a symbol of idealism, courage and unselfishness.

Greece 1821–29

1821 An uprising against Turkish rule is countered with hard resist-ance and great brutality. Thou-sands of Greeks are massacred or sold into slavery.

1827 An allied fleet of British, French and Russian naval forces annihilate the Turkish-Egyptian fleet at Navarino. Thus Turkey is forced to give up Greece.

1828 The former Russian foreign minister, Count Kapodistrias, becomes the country's interim president.

1830 Great Britain, France and Russia recognize Greece as an independent state.

Russian Expansion
1689–1860 (map nr. 126, p. 89)

1689–1725 The reign of Peter the Great (see p. 71).

1762–96 Catherine the Great makes Russia a European great-power (see p. 73). Rebellion led by the Cossack Pugachev (see p. 83).

1814–15 Following the Congress of Vienna, Russia gets three fourths of the Grand Duchy of Warsaw (see p. 85). Tzar Alexander I (see p. 83) calls for the establishment of a «Holy Alliance», directed against the forces of revolution and libera-lism. When the alliance is formally established in September 1815, most of Europe's princes join the cause.

1816–17 Serfdom is abolished in the Baltic provinces, but the peas-ants receive no land, remaining dependent on wealthy estate owners.

1825 Alexander I dies with no heir at the age of forty-seven, and his brother, Constantine Pavlovitch, viceroy of Poland, is proclaimed Tzar. However, Constantine has two years earlier renounced his claim to the throne in favour of his younger brother Nicholas. During the period of uncertainty that re-sults, a secret society of officers demands a new and more liberal constitution and system of govern-ment. In St. Petersburg they refuse

A Russian girl carries buckets of water with a yoke. Ca. 1850.

Russian peasants dancing a ring dance in the 1850s.

A soldier in Tzar Nicholas I's imperial guard.

to swear loyalty to Nicholas, proclaiming Constantine instead. The coronation of Nicholas I takes place in December (Russian: *decabr*), and the rebels are therefore referred to as the *Decabrists*. Nicholas I orders the use of artillery against them, and the Decabrist revolt is quickly crushed.

1825–31 Nicholas I's reactionary policies are met with growing opposition. A war with Iran makes portions of Armenia Russian. The country gives aid to the war of liberation in Greece in 1828. Poland becomes a Russian province following a Polish uprising in 1830–31.

1853–56 Crimean War (see column 4).

126. RUSSIAN EXPANSION 1689-1860

▨ Russian territory in 1689; real control in the east only in patches

Conquests:
▢ 1699-1725   ▢ 1725-1800
▢ 1800-1815   ▢ 1815-1860

For Western Russia see also maps 101 and 120

Below: *An allegorical colour print offers a satirical explanation for the Crimean War: The Tsarist eagle is putting Turkey in its cage.*

Below right: *The Scottish Guard charges at Alma in the first battle of the Crimean War, September 20, 1854. The allies – Turkey, France and Great Britain – had only limited forces at their disposal, but defeated the Russians all the same.*

Left: *The promenade along the Neva in St. Petersburg, Russia's capital 1712–1918. From a painting by Carl Beggrow from the middle of the 1800s. In the foreground is the old Hermitage Museum, built 1771–81, in the background the Hermitage Theatre from 1783–87.*

The Crimean War 1853–56

In 1853 the Russians moved into Moldavia and Wallachia (map nr. 141, p. 101). Turkey declared war on Russia in October, and in March of 1854, Great Britain and France entered the war on the side of Turkey. The war was the first to get broad press coverage, and one figure to receive well deserved notice was the British nurse, Florence Nightingale for her contribution. Russia had to capitulate, and March 30, 1856 a peace was reached in Paris.

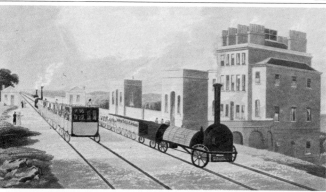

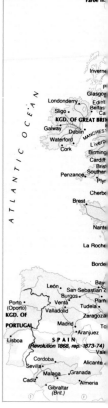

The Railways

must be seen as one of the most revolutionary innovations of the 19th century. But like so many other inventions, it too had an evolution. For a long time it had been common to use wagons pulled on tracks of wood or iron in the mines, and in the 1770s the Frenchman Joseph Cugnot had constructed a steam-driven wagon. In England, George Stephenson combined these two systems, and in 1814 he had constructed a steam-driven wagon that could pull eight ore cars (see illus. p. 77). The invention was at first not an unqualified success. However, in 1825 Stephenson demonstrated a new locomotive, and showed that it was the transportation system of the future. This took place on a short stretch of track between Stockton and Darlington.

The picture at the top of the page shows the first trains on the Liverpool-Manchester run, which was opened in 1830. During the construction phase a contest had been announced for builders of locomotives. And in the famous trial at Rainhill, George Stephenson won the prize with his locomotive, «The Rocket».

Freight transport on the canal between Liverpool and Manchester took thirty-six hours; the stage coach could cover the distance in 4 1/2 hours. «The Rocket» covered the same distance in under two hours!

In the beginning only short, point-to-point lines were built, but from about 1860, Great Britain, Belgium and some German states began construction of interconnected railway systems. Fifteen years later areas of Austria-Hungary, Italy, Spain and France were also linked to the railway systems in the German Empire, Russia and Romania (map nr. 127).

In 1830 Europe had 316 km of tracks, 279 km of these in Great Britain. In 1865 this had increased to a total of 75 882 km, and ten years later, 142 494 km.

The picture below, from Charles Rossiter's painting «Brighton and Back for 3/6» from 1859, shows a crowded, but inexpensive third class car. Railroad cars of the time had three classes. The least expensive were used by the petty bourgeoisie and the working class.

Right: The world's largest iron-hulled ship at the end of the 1850s, the «Great Eastern», drawn and constructed by British engineer Isambard K. Brunel. Construction of the 207 m long ship – driven by paddle-wheel, propeller and sail – was begun in 1852. September 9, 1859 marked her maiden voyage, which, however, ended in tragedy. An overheated boiler blew up, killing six machinists, and destroying the bow section of the ship. Brunel himself died of a heart attack only days later. The «Great Eastern», which was built to carry 4 000 passengers on voyages to India and Australia, was an economic fiasco. Because of the vessel's huge size, it was difficult to fill. Nor did it help matters that the ship was extremely unstable at sea.

American Robert Fulton's steamship «Clermont» from 1807.

Poor children in one of London's slum districts in the 1860s.

Women's liberation. German caricature from 1848.

127. EUROPE 1870. DENSITY OF POPULATION, COMMUNICATIONS

Inhabitants per square kilometre:

▨ More than 100

▨ 20-100

▨ Less than 20

■ City with more than 1 million inh.: PARIS

• City with 500 000-1 million inh.

⌣⌣⌣⌣⌣ Important canal

— Railway built 1827-70 (the most important in British Isles and C. Europe are shown).

— Frontier

See also map 107

Changes in Population

As seen from the map at the left, there were only four cities in Europe in 1870 with more than a million inhabitants: London and Manchester in Great Britain, Paris and Constantinople (Istanbul).

In Europe – which had by far the greatest population density of all the continents – it was only in parts of Great Britain and Italy, and in areas of Central Europe, that the average density was greater than 100 inhabitants per sq. km.

Due to emigration to the Americas and Australia at the end of the 1800s, the populations of these areas were growing faster than that of Europe. Research has shown that on average the population of Europe grew by 50 percent, 90 percent in South and Central America, 220 percent in North America, and a whopping 300 percent in Australia and Oceania.

American paddle-wheeler on the Mississippi in the 1860s.

Transporting bails of cotton from a plantation in the Southern States.

George Washington is inaugurated president in New York on 31 April 1789.

Above: The so-called «Boston Massacre», May 5, 1770. From a contemporary copperplate engraving by Paul Revere. Like other illustrators of his day, Revere has dramatized the event in order to strengthen opposition to the British. It began with some youngsters throwing snowballs at two British sentries. The sergeant of the guard was summoned, several adults arrived, a soldier was pushed to the ground, and suddenly shots rang out – five colonists were killed and six wounded. To increase the effect of the picture, Revere has written Butchers Hall over the entrance to the British customs house.

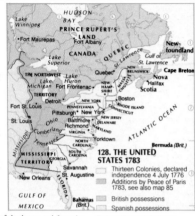

128. THE UNITED STATES 1783

Thirteen Colonies, declared independence 4 July 1776

Additions by Peace of Paris 1783, see also map 85

British possessions

Spanish possessions

The United States Of America (USA) 1783–1867

1783 The final peace treaty following the American war of independence is signed in Paris. The independence of the colonies is recognized, and their territories are expanded (map nr. 128).

1787 A constitutional convention in Philadelphia works out a new constitution.

1789 The constitution is in effect. George Washington becomes the union's first president (see illus. below).

1792 The political parties are formed. Alexander Hamilton and John Adams are in favour of a strong federal state and form the Federalist Party. Thomas Jefferson, an avid opponent of centralization, founds the Democratic-Republican Party (later the Democratic Party).

1801 Thomas Jefferson takes the oath of office in the new capital of Washington.

1803 The USA purchases Louisiana from France for 80 million francs (map nr. 130, p. 93).

1812–14 War with Great Britain. British Canada's friendly attitude to the indians, who oppose the American government and clash regularly with American settlers moving westward, is one of the reasons for the war. In 1814 the

Below: George Washington (1732–99) was commander of the colonial army from 1775 in the American war of independence, and made such a favourable impression – in Europe as well – that Frederick II of Prussia sent him his portrait with the dedication: «From Europe's oldest general to the world's greatest». After serving as chairman of the committee that was to write a new constitution for the union, Washington was elected first president of the United States of America in 1789. He was reelected in 1792, but firmly refused to run for a third term. This portrait of Washington is from James Sharples's contemporary painting.

129. THE AMERICAN CIVIL WAR 1861-65

States in the Union (California and Oregon on west coast)

Slave states in the Union

S. Carolina out of Union, Dec. 1860

States that followed South Carolina:

Jan-Feb 1861   April-May –61

Sep. fr. Virginia 1861, Union 1863

Boundary Union - Confederacy

Slave auction in the Southern States in the mid-1800s.

Indian lodge from the 1790s, built of logs.

General George Meade leading the Northern army at Gettysburg in 1863.

Above: The first shots of the American Civil War were fired by Confederate forces from Charleston, South Carolina at Fort Sumter, which was situated on an island in the harbour, and manned by ten Union officers and sixty-five men. The attack began at dawn on April 12, 1861, and the next afternoon – after 3000 shells had been fired at the fort – the company surrendered. Only one soldier was killed at Fort Sumter, but in the four years the Civil War raged, 650000 Americans lost their lives, out of a total population of 30 million. This is approximately 150000 more than America lost during the Second World War.

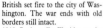

British set fire to the city of Washington. The war ends with old borders still intact.

1819 Spain sells Florida to the USA for five million dollars.

1823 President James Monroe issues the so-called Monroe Doctrine, which states that the USA will consider any attempt on the part of European states to colonize or otherwise interfere in the affairs of the North American continent to be an act of aggression.

1846–48 War with Mexico. Mexico cedes New Mexico and California to the USA.

1860–61 Abraham Lincoln, who

has run on a platform that would abolish slavery, is elected president. Six weeks later – in December of 1860 – South Carolina leaves the union. In the course of the spring, ten other states secede from the union. The eleven states that have seceded form their own union, The Confederate States of America.

1861–65 On April 12 the first shots are fired (see illus. above). In the early years of the war the Southern States are better armed and win important victories, but the North-

ern States have far greater resources. In September 1862 the northern army wins a key victory at Antietam. The following year general George Meade defeats the famous general of the Southern States, Robert E. Lee, at the Battle of Gettysburg. April 9, 1865 Lee surrenders to the Northern States' Ulysses S. Grant at Appomattox.

1867 The USA purchases Alaska from Russia for 7.5 million dollars.

130. THE UNITED STATES 1783-1912

▨ The Thirteen States in 1776
☐ Ceded by Britain 1783
☐ Spain to France 1800, sold to U.S. 1803
☐ Independence from Mexico 1836, annexed by U.S. 1845
▨ Incorporated 1846
▨ Ceded by Mexico 1848
☐ United States (Union) 1861-65
☐ Confederacy 1861-65
— First transcontinental railway, completed 10 May 1869
Dates are year when state was incorporated in the union

A partially veiled peasant girl from South America in the 1830s.

Sout American indians around 1830.

A Mexican milit. policeman in Pre dent Diaz' guard

Liberation of Spanish America

Beginning in about 1810, Simon Bolivar (1783–1830) was actively engaged in the Latin American struggle for independence. He joined a group in New Granada (map nr. 131), and in 1819 he was made president of Venezuela – the eastern part of New Granada. The same year the western part was also incorporated, and the country of

«Gran Colombia» proclaimed (map nr. 132), with Bolivar as its president. In 1822 he liberated present-day Ecuador, which also became a part of Gran Colombia. He then took part in Peru's struggle for liberation, and the northern part of the country - Bolivia - is named after him. In 1824 Spanish hegemony in South America came to an end. **The picture** of Bolivar, **top left,** is from a contemporary painting.

United States Intervention in Cuba 1898

Because the United States wanted to strengthen its Pacific fleet in the 1890s, it wanted to take on the construction and administration of the Panama Canal. To accomplish this, the U.S. would also have to control the Caribbean. Just such an opportunity presented itself in 1895 when Cuba rebelled against Spanish rule. There was an imme-

diate cry to «come to the aid of Cuba,» and in 1898 the U.S. declared war on Spain. Within a few months the Spanish had been driven out of Cuba. **The picture below** shows Theodore (Teddy) Roosevelt, leading his «Rough Riders». In 1900 he became vice president, and following the assassination of president William McKinley in 1901, Roosevelt became president.

Argentinian «gau-chos», drovers, at the beginning of the 1800s.

 A female Mexican revolutionary soldier from the end of the 1800s.

 French officers in Mexico in 1863.

Mexico 1519–1911

1519–21 N. part of Spanish America, the Viceroyalty of New Spain (map nr. 131) is taken from the Aztecs and Toltecs (see p. 63).

17th and 18th centuries Growing unrest between Spanish civil servants and the Spanish colonists.

1810–24 The War of Independence against the Spanish begins in 1810. In 1821 Augustin de Iturbide proclaims himself emperor, but has to step down in 1823. The following year Mexico becomes a federal republic.

1845–58 Following a war of liberation Texas declares its independence, and is incorporated by the U.S. in 1845 (map nr. 130, p. 93). American activities in the northwestern territories lead to war with the U.S. The Americans take Mexico City in 1848. California becomes an American state in 1850, and three years later the Americans purchase a small area in present-day Arizona (map nr. 130, p. 93). The indian Benito Juarez becomes president of Mexico in 1858.

1861–67 Sends French troops to Mexico. The Austrian archduke Maximilian is installed as emperor in 1864, but when the French have withdrawn, he is shot on June 19, 1867. Juarez becomes the country's leader again, and remains so until his death in 1872.

1877–1911 President Porfirio Diaz – who in the **picture top right on the previous page** is seen in the company of money-grubbing captains of industry during a dance recital – is supported by the army and the estate owners, and gives foreign capital a free hand. Following a revolution in 1911, Diaz goes into exile in Paris.

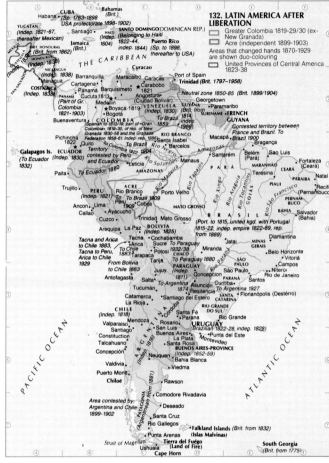

132. LATIN AMERICA AFTER LIBERATION

Greater Colombia 1819-29/30 (ex-New Granada)

Acre (independent 1899-1903)

Areas that changed hands 1870-1929 are shown duo-colouring

United Provinces of Central America 1823-38

Argentina 1776–1842

1776 Spain establishes the viceroyalty of Rio de la Plata (map nr. 131, p. 94).

1806–07 A British force of 10 000 men occupies Montevideo and advances toward Buenos Aires. They are stopped and forced to withdraw.

1810 The province of Buenos Aires refuses to submit to Spanish rule.

1816 The province of Rio de la Plata declares its independence.

1826 Argentina a republic.

1842 Spain recognizes the country's independence.

Brazil 1549–1822

1549 A Portuguese colony since 1500, gets governor general. Bahia (Sao Salvador) becomes governor's residence. The Jesuits begin missionizing among the indians.

1750–77 The Jesuits are driven from the country. Rio de Janeiro becomes the new capital in 1762.

1815 Independent monarchy, governed by Portuguese crown regent John VI, whom Napoleon has exiled.

1822 John's son, Pedro I, becomes emperor of an independent Brazil.

Right: German naturalist and geographer Alexander von Humboldt (1769–1859). From a contemporary painting. From the summer of 1799 to early 1804 Humboldt made an epoch-making journey in South America together with Aimé Bonpland. And in 1805 started work on a third volume work, «Voyage aux régions équinoxiales du nouveau continent», which it took thirty years to complete. Humboldt undertook several voyages of exploration later, and there are over a thousand geographic locations around the world that he named.

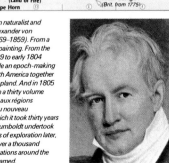

British cavalryman in wood, made by an African in the 1800s.

Woman with child is put on the auction block in Marrakesh in 1905.

Livingstone (at left, and Stanley meet in Ujiji in the spring of 1871.

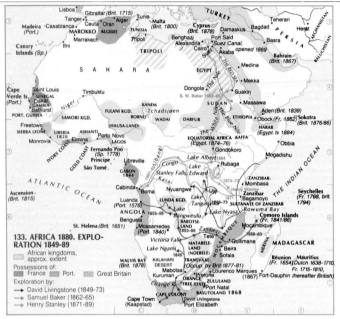

133. AFRICA 1880. EXPLORATION 1849-89

African kingdoms, approx. extent

Possessions of:
France   Port.   Great Britain

Exploration by:
→ David Livingstone (1849-73)
→ Samuel Baker (1862-65)
→ Henry Stanley (1871-89)

above shows his next two expeditions as well (1858–64) and (1866–73). It was during this last journey, to the area around Lake Tanganyika in search of the source of the Nile, that contact with Livingstone was lost. In 1869 the English-born American journalist Henry Morton Stanley (1841–1904) was ordered by his editor in chief to find Livingstone, «no matter what the cost.» In March 1871 he started out from Zanzibar, and in early November Stanley reached Ujiji. It was there he caught sight of a man who looked as though he could be Livingstone and uttered the line of the century: «Dr. Livingstone, I presume.» The missionary's strength was failing, and he died on May 1, 1873.

Stanley, who remained with Livingstone for four or five months, later explored the upper Congo River, founded the Free State of the Congo and stayed on as its governor. The portrait of him at the **top of the page** was painted by his wife in 1893.

The Nations of the Barbary Coast,

or the **Pirate States**, as they were also known, was an older name for Morocco, Algeria, Tunisia and Libya (map nr. 103, p. 72). They made their living, quite simply, through piracy, and the tribute paid

134. FRENCH CONQUEST OF ALGERIA 1830-1902

Conquests in:
1830-35   1871-1900
1836-47   After 1900
1848-70   Present-day boundary

for safe passage. This continued until the French drove the Turks from the area.

Algeria was conquered in the 1830s, but as the map below indicates, it was some time before the French gained control over the whole country. **Tunisia** was a French protectorate from 1881, and **Morocco** was slowly subjugated by the French in the period 1906–34. **Libya** became an Italian colony in 1912.

The influx of Europeans was great, especially in Algeria, where in 1906 there were already 450 000 Frenchmen.

Livingstone and Stanley

The Scottish physician and missionary David Livingstone (1813–73) travelled as a missionary at age twenty-seven to Southern Africa.

In 1841 he reached Bechuanaland, where he practised as doctor and missionary for eight years before beginning his long trek northward. He established a number of mission stations and began his struggle against the extensive slave trade in the interior of Africa.

In 1852 he sent his family home and set off alone on his first great exploration westward. The map

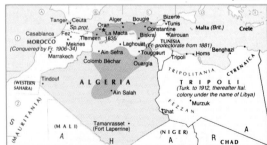

African warrior from the 1800s.

The «Boers» move inland during «The Great Trek» (see below).

President Krüger ready for battle. Caricature from 1899.

135. SOUTHERN AFRICA 1899-1910

▨ British colony or protectorate 1899

▨ Boer republic

— Union of South Africa, 1910

★ The Great Trek, 1836

★ Battlefield

Dates indicate year of British occupation

Sudan Becomes ish

northern Sudan was occupied ypt in 1821. In spite of Afri- resistance, Egyptians and eans carried on an extensive trade in the years that follo- But in 1881 the Islamic people r the Mahdi rebelled and d their foreign rulers. At the of the century British and tian troops advanced south- to retake the Sudan. **The ure below** shows a scene from lecisive battle at Omdurman, ember 2, 1898. Ca. 20 000 ishmen, led by general Her- Kitchener, and armed with ern weapons, including machi- ns that could shoot nearly 700 ds per minute, met 50 000 l warriors, armed with spears old rifles. In just a few short s, the «executions» were over. nd half of the natives had been down, with 11 000 dead. The sh lost 386 men.

The Boers and the Boer War 1899–1902

After the Cape Colony had changed hands from Dutch to British in 1806, the white farmers, who in Dutch were called «Boers», began to leave. In 1836 began the «Great Trek» northeastward into the inte- rior. There they established the two Boer republics: Transvaal in 1852, and the Orange Free State in 1854 (map nr. 135).

The British hoped to include the Boers in a South African union under British dominion, but the Boers were not interested. The rich gold deposits discovered in Trans- vaal in 1886 did nothing to reduce the British appetite for the territo- ries. In the autumn of 1899 the British increased troop strength in South Africa, and when they rejec- ted an ultimatum from Transvaal president Paul Krer to withdraw these forces, the Boers attacked.

Both Mafeking and Kimberley in British Bechuanaland were besieg- ed. Great Britain sent large num- bers of reinforcements to South Africa, and in the summer of 1900 the British took Johannesburg, Pretoria and Bloemfontain. Some months later the two states

were annexed, but the Boers conti- nued a guerilla war that led the British to take very unconventional countermeasures. They established concentration camps in which conditions were so appalling that it even aroused disgust at home in Britain. As a condition of the peace in 1902, the Boers were forced to recognize British hegemony.

The picture at the top of the page shows horses, pulling a canon into position during the war. From a contemporary painting by George Scott.

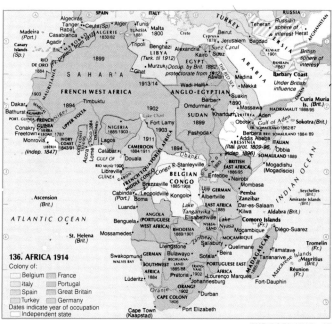

136. AFRICA 1914

Colony of:
▨ Belgium ▨ France
▨ Italy ▨ Portugal
▨ Spain ▨ Great Britain
▨ Turkey ▨ Germany
▨ Independent state
Dates indicate year of occupation

Little girl from the mid-1800s, fashionably dressed.

Two British soldiers during a parade for Queen Victoria.

Unemployed British worker around 185...

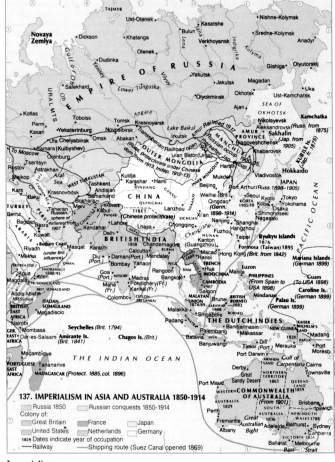

137. IMPERIALISM IN ASIA AND AUSTRALIA 1850-1914

- Russia 1850
- Russian conquests 1850-1914

Colony of:
- Great Britain
- United States
- France
- Netherlands
- Japan
- Germany

1859 Dates indicate year of occupation
— Railway
— Shipping route (Suez Canal opened 1869)

Two of the British Empire's foremost figures. **Top:** *Queen Victoria, reproduced from a portrait in «The Times» on the day that marked her 60th year on the British throne, June 21, 1897. Not only did Victoria lend her name to the zenith of British imperial history, but also to the period style in architecture, art and literature. She was a strong-willed regent, who often protested against the decisions of her prime ministers. This was especially bothersome for William Ewart Gladstone, seen* **below.** *He served four terms as prime minister between 1864 and 1894.*

Imperialism

The maps on this placard illustrate the colossal European expansion in other parts of the world at the end of the 19th century.

At the time of Queen Victoria's diamond anniversary, for example, the British Empire included a quarter of the world's population. It was four times larger than the Roman Empire had been.

138. THE BRITISH EMPIRE 1815-1914

- 1815
- 1914

Colonies in 1815 lost before 1914 are not shown

British troops employed elephants during battles in India.

Disraeli hands Victoria the imperial crown of India.

India got its first railway in the 1850s, constructed by the British.

139. EUROPEAN POSSESSIONS IN SOUTHEAST ASIA 1824
- British
- Portuguese
- Dutch
- Spanish

•Canton
•Macao (Port.)

PACIFIC OCEAN

BURMA
•Rangoon
SIAM
Ayutthaya
•Hué
ANNAM (VIETNAM)
LAOS
Luzon
Nicobar Is. (Danish 1756-73 and 1846-56)
•Bangkok
•Angkor
COCHIN CHINA
Manila
Gulf of Siam
•Saigon
PHILIPPINES (Spanish fro 1565, Brit. 1762-64, thereafter Spanish to 1898)
Palawan
Sulu Sea
Mindanao
SOUTH CHINA SEA
•Patani
•Atjeh
•Penang(Brit. from 1786)
MALAYA
Brunei
SABAH
Natun Is.
Celebes Sea
Halmahera
Malakka (Brit. from 1824)
SARAWAK
•Singapore (Brit. from 1819)
•Sambas (Brit. 1795-1824)
Manado
Nias
Pontianak
BORNEO (KALIMANTAN)
MOLUCCAS
Padang
Sukadana
Celebes (Sulawesi)
Sula
Ceram
Palembang
Bangka
•Amboina
Benkulen•
Billiton
Bandjarmasin (Brit. 1812-16)
Buru
Banda Sea
Bantam•
Batavia
Jakarta
Java Sea
Makassar
JAVA
Surakarta
Surabaja
Flores Sea
Tanimbar Is.
Jogjakarta (Mataram)
Bali
Sumbawa
Timor
THE INDIAN OCEAN
Lombok
Sumba
Flores

Top right: Bombay's magnificent and luxuriously decorated Victoria Station from 1877–87. Looking more like a cathedral than the headquarters of the Indian Railways, it is an impressive symbol of British hegemony in India.

India 1756–1885

1756–63 Robert Clive drives the French out of India. The British East India Company takes power in many parts of the country.

1857–59 An Indian uprising, the Sepoy Rebellion (known to Indians as the First War of Liberation), is put down. The administration of India is transferred from the East India Company to the British Crown.

1877 India becomes an empire, with Victoria as empress.

1885 The Indian Congress Party is formed and advances the struggle for India's liberation.

140. THE COLONISATION OF INDIA 1765-1914
- British colonies in 1765
- Additions 1765-1858
- British colonies in 1914
- British subject states 1914, under Viceroy or Indian administration
- Boundary of British India 1914 (Indian Empire from 1877)
- **1883** Dates indicate year of colonisation

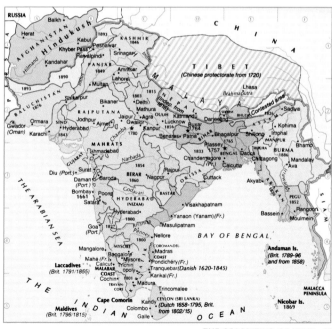

RUSSIA
Balkh•
AFGHANISTAN
Hindukush
Herat
Kabul
KASHMIR 1846
Khyber Pass
Peshawar
Srinagar
1893
Himand
Kandahar
Rawalpindi
SIKKIM
1826
BHUTAN
CHINA
TIBET (Chinese protectorate from 1720)
Lhasa
Brahmaputra
Sadiya
Contested area
PANJAB 1849
•Multan
Lahore
•Amritsar
Indus from 1843
BALUCHISTAN
1876
•Shikarpur
Bikaner
•Delhi
1803
1815
Mathura
Kathmandu
NEPAL
Darjeeling
ASSAM 1826
•Kohima
Gwador (Oman)
Ormara
SIND
Jodhpur
RAJPUTANA
Jaipur
Ajmer
•Agra
OUDH
•Lucknow 1856
Imphal
MANIPUR
Karachi 1843
•Hyderabad
Gwalior
1780
Kanpur
Buxar 1764
Benares•
Patna•
BIHAR
Bhagalpur
Shillong
TRIPURA 1886
BURMA
Bhamo
GUJARAT
MAHRATS
Ahmadabad
Chamba
Narbada
1833
Plassey 1757
BENGAL
Dacca
Chittagong
Mandalay
•Ava
Diu (Port.)
Surat
BERAR 1860
Rajpur
Chandernagore (Fr.)
Calcutta
Akyab
Daman (Port.)
•Baroda 1661
Bombay•
•Poona
Godavari
ORISSA
Cuttack
PEGU 1852
Satara
BASTAR
HYDERABAD (NIZAM)
Rangoon
•Hyderabad 1800
SIAM
Bassein•
Moulmein
Goa (Port.)
1817
Penner
•Nellore
•Visakhapatnam
THE ARABIAN SEA
•Masulipatnam
Yanaon (Yanam)(Fr.)
BAY OF BENGAL
Mangalore•
MYSORE
COROMANDEL COAST
•Madras
Andaman Is. (Brit. 1789-96 and from 1858)
Mahé (Fr.)
•Bangalore
•Mysore
Laccadives (Brit. 1791/1855)
Calicut•
Trichin opoly
MALABAR COAST
•Pondichéry(Fr.)
Tranquebar(Danish 1620-1845)
Cochin•
1801
•Karikal (Fr.)
TRAVAN-CORE
•Madura
Trincomalee
MALACCA PENINSULA
Cape Comorin
CEYLON (SRI LANKA)
Kandy
•Colombo
Nicobar Is. 1869
Maldives (Brit. 1796/1815)
Galle•
(Dutch 1658-1795, Brit. from 1802/15)
THE INDIAN OCEAN

Many Russian soldiers froze to death during the campaign in the Balkans.

A flower girl in London in 1888.

Two telegraphists at the main telegraph station in London in 1872.

Above: Otto von Bismarck (1815–98). From a contemporary painting by Franz von Lenbach. Prince Bismarck, who created the first unified German state in 1871, held on to his position of power for almost twenty years.

Below: November 21, 1783 marked the first flight of an untethered hot-air balloon with passengers on board. The event, reproduced in this lithograph, took place in the Boulogne Forest in Paris. It was the brothers Joseph and Étienne Montgolfier who had constructed the balloon, later known as the Montgolfier.

Unrest in the Christian States of the Baltic in the 1870s

In the summer of 1875 the people of Herzegovina rebelled against their Turkish overlords. The unrest soon spread to Bosnia, and in the spring of 1876 the Bulgarians joined the revolt. In response the Turks murdered in all 12 000 of the country's men, women, and children.

On June 30 Serbia declared war on Turkey, and a few days later Montenegro followed suit. By September 1 they had been defeated by Osman Pasha's forces. The following week William Gladstone wrote his sharp and angry pamphlet: «The Bulgarian Horrors and the Question of the East».

Tzar Alexander II decided to come to the rescue of the Balkan states, but first he let Austria know he would not oppose an Austrian occupation of Bosnia and Herzegovina. In April 1877 Russia declared war on Turkey.

Surprisingly, on the way to Constantinople the Russians were stopped by the Turks near Plevna. 14 000 Turks held off a superior Russian-Bulgarian force from July 20 to December 10, 1877.

After a hard winter in the Balkan Mountains the Russians were able to make it over the Shipka Pass and advance on Constantinople.

Then, on February 15, 1878, divisi-

141. SOUTHEASTERN EUROPE AFTER THE CONGRESS OF BERLIN 1878

— Ottoman Empire's boundary before prel. Peace of San Stefano in March 1878

--- Northern boundary of Ottoman tributary states

Ottoman Empire after Congress of Berlin

Greater Bulgaria after preliminary Peace of San Stefano in March 1878

Occupied by Austria-Hungary 1878

...... Boundaries before 1878. Areas duo-coloured changed hands by Congress of Berlin

ons of the British navy arrived in Constantinople, which meant that the Russian supply lines could now be interrupted at short notice. Both the Turks and the Russians now preferred to make peace. It was concluded in San Stephano in March 1878. At the insistence of Russia, one result of the war was the establishment of Greater Bulgaria (map nr. 141). The fact that this gave Russia access to the Aegean was equally troubling to both Great Britain and Austria. Bismarck took on the role of mediator and summoned all the relevant parties to

The Congress of Berlin

It was assembled for a month, beginning June 13 1878, and resulted in Greater Bulgaria being divided in three parts. Macedonia was once again Turkish, and Bulgaria and East Rumelia were under Turkish dominion. The Russians got Bessarabia and Batum. Yet, the fact that Austria's occupation of Bosnia and Herzegovina was accepted, sowed the seeds of the conflict that would throw the entire world into war in the summer of 1914 (see p. 103).

A tennis match at Wimbledon in 1881.

A woman's body could be badly mistreated by the corsets of the day.

Female tourist in a sedan chair on her way to the top of Vesuvius, ca. 1880.

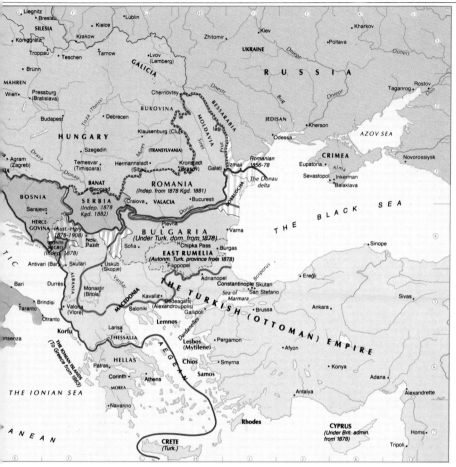

Liegnitz
Breslau
SILESIA
Königgrätz
Troppau
Teschen
Brünn
MÄHREN
Wien
Pressburg
(Bratislava)
Kielce
Lublin
Krakow
Tarnow
GALICIA
Lvov
(Lemberg)
Chernovtsy
BUKOVINA
Klausenburg (Cluj)
Debrecen
Szegedin
HUNGARY
Budapest
Agram
(Zagreb)
Temesvar
(Timisoara)
BANAT
Beograd
(TRANSYLVANIA)
Hermannstadt
(Sibiu)
Kronstadt
(Brasov)
MOLDAVIA
Iasi
Izmail
Galati
Romanian
1856-78
The Donau
delta
DOBRUDJA
ROMANIA
(Indep. from 1878 Kgd. 1881)
Craiova
VALACIA
Bucuresti
Donau
BOSNIA
Sarajevo
SERBIA
(Indep. 1878
Kgd. 1882)
HERCE-
GOVINA
(Aust.-Hgry.
1878-1908)
Novi
Pazar
MONTE-
NEGRO
(Indep. 1878)
Sofia
Plevna
BULGARIA
(Under Turk. dom. from 1878)
Chipka Pass
Varna
Burgas
EAST RUMELIA
(Autonm. Turk. province from 1878)
Filippopel
Zhitomir
Kiev
UKRAINE
Kharkov
Poltava
R U S S I A
Dnjepr
Dnjestr
Bug
Kherson
Odessa
JEDISAN
AZOV SEA
Rostov
Taganrog
Don
CRIMEA
Novorossiysk
Eupatoria
Alma
Sevastopol
Inkerman
Balaklava
THE BLACK SEA
Sinope
Ereğli
Antivari (Bar)
Skutari
Uskub
(Skopje)
ALBANIA
Monastir
(Bitola)
Durrës
Brindisi
Taranto
Otranto
Valona
(Vlore)
Korfu
THE IONIAN ISLANDS
(To Greece from 1863)
Larisa
THESSALIA
HELLAS
Patras
MOREA
Corinth
Athens
Navarino
THE IONIAN SEA
Adrianopel
Kavalla
Dedeagatsj
(Alexandroupolis)
Saloniki
MACEDONIA
Vardar
Lemnos
Gallipoli
Dardanelles
Lesbos
(Mytilene)
Chios
Samos
Pergamon
Smyrna
Constantinople Skutari
San Stefano
Sea of
Marmara
Bosporus
Brussa
Ankara
Sivas
THE TURKISH (OTTOMAN) EMPIRE
A E G E A N
Rhodes
CRETE
(Turk.)
Afyon
Konya
Antalya
Adana
CYPRUS
(Under Brit. admin.
from 1878)
Alexandrette
Homs
Tripoli
ADRIATIC
Bari

Right: The foremost statesmen in Europe met in 1878 at the Congress of Berlin, led by Otto von Bismarck, seen here sitting with his back to the centre window. In front of the window at the left sits the British foreign minister, Lord Salisbury, and Prime Minister Benjamin Disraeli, who together with Bismarck played the most important role in the congress. The Russian delegation is sitting at the end of the table on the left, with the Turks opposite at the table on the right. Members of the delegation from Austria-Hungary are sitting on Bismarck's right, and the French on his left.

THE CONGRESS OF BERLIN 101

A Japanese soldier from the time of the Russo-Japanese War.

 A Japanese geisha from the late 1700s.

 Japanese rickshaw, probably invented around 1870.

142. JAPANESE EXPANSION 1868–1939

Japan in 1868
Conquests:
1868-05   1905-10   1931-39
Dates indicate year of occupation
— Railway

Sidetrack finished 1917

Sakhalin (Karafuto) (1905)

Kurile Islands (1875)

Amur
Khabarovsk

MANCHURIA (MANSHUKUO) (Japanese 1932-45)

Hokkaido

Otaru

OUTER MONGOLIA (Indep. 1912/13)

Harbin

Songhuajiang

Trans-Siberian railroad (completed 1904)

Vladivostok

Hakodate

Akita

JILIN

Trans-Siberian railroad completed 1904

INNER MONGOLIA (1936)

Changkufeng

JAPAN

The Great Wall of China

Mukden
LIAONING

Railroad completed 1935

LIAODONG

Pyongyang

Sendai

SEA OF JAPAN

Niigata

Beijing (Peking) (1937)
Tianjin (Tientsin)   Tanggu

Yalu

Dairen
Port Arthur (1905)

BOHAI

Seoul (Keijo)

Kanazawa

Tokyo (Edo)
Yokohama

Honshu

SHANNI

SHANDONG

Weihai

KOREA (CHOSEN) (1910)

Kyoto
Kobe

Nagoya
Osaka

HEBEI

Huanghe

Qingdao

THE YELLOW SEA

Pusan (Fusan)
Mokpo

Tsushima

Shimonoseki

Shikoku

Yan'an

Kaifeng

JIANGSU

Cheyu-do

Strait of Tsushima

Yahata
Nagasaki

Kyushu

Xi'an

SHAANXI   HENAN   ANHUI

EAST CHINA SEA

Kagoshima

PACIFIC OCEAN

CHINA

Nanjing (Nanking) (1937)   Shanghai (1937)

HUBEI

Hangzhou

SICHUAN

Hankou (1938)

Changjiang

Chongqing (Chungking)

Yongjia

JIANGXI

Wenzhou

Ryukyu Islands (1876)

GUIZHOU   HUNAN

FUJIAN

Fuzhou

Taiwan (Formosa) (1895)

GUANGXI – ZHUANGZU

GUANGDONG

Xiamen (Amoy) (1938)

Shantou (Swatow)

Nanning

Guangzhou (Kanton)

Macao (Port. 1557)   Hong Kong (Brit. 1842)

ZHUANGZU

Peihai (Pakhoi)   Zhanjiang

Haikou

SOUTH CHINA SEA

Hainan (1939)

THE PHILIPPINES

Luzon

143. RUSSO-JAPANESE WAR 1904-05

Occupied by Russia 1900-05
Japanese thrust   — Railway

INNER MONGOLIA

Harbin (Pinkiang)

MANCHURIA

RUSSIA

Vladivostok

REHE

CHINA

Mukden
Laoyang

Chongin

KOREA (Jap. protectorate)

Yalu

Linyo

LIAODONG

Dalian
Port Arthur (Russian 1898-1905)

Andong
Pyongyang

SHANDONG   Japanese 1905-45)

Weihai (Brit. 1898-1930)

Chinnanpo

Kaesong

Inchon (Chemulpo)

Seoul (Keijo)

Qingdao (German 1898-1914)

SEA OF JAPAN

Tsushima 27.5 1905

JAPAN

Japan 1867–1939

1867–69 The Shogunate is abolished and replaced by a strong imperial system with Meiji Mutsuhito (1852–1912) on the throne. End of feudalism. The Damyos (noblemen) voluntarily transfer their territories to the emperor.

1875–76 Russia recognizes Japan's rights to the Kuriles (1875) and the Ryukyu Islands (1876).

1889 New constitution, based on Prussian model, but real power is retained by the emperor.

1894–95 China forced to cede Taiwan (Formosa).

1904–05 Japan declares war on Russia February 10, 1904. Russians suffer enormous losses. Following the battle in the Strait of Tsushima (illus. below), Russia has to capitulate. Japan gains hegemony over Port Arthur and Sakhalin.

1910 Korea (Chosen) becomes Japanese province.

1914–18 During WW1, the country takes part on the side of the Entente powers (Great Britain/ France).

1919 At Versailles, Japan receives the German possessions on the Shandong Peninsula, and the German islands in the Pacific that Japan occupied during the war. Occupied territories in China and Russia must be abandoned.

1931–33 Manchuria is taken and made the Protectorate of Mandshukuo (1931–32). When the League of Nations condemns the action, Japan withdraws its membership in the League (1933).

1936 Japan signs anti-Comintern pact with Germany. Later, Italy also joins the pact.

1937 Japan attacks China again, with great success, but in spite of military superiority, Japan is unable to crush Chinese resistance.

1939 The USA revokes its trade agreement with Japan.

Right: A good year after the outbreak of the Russo-Japanese War – May 27, 1905 – the Russian fleet, having sailed halfway round the world, was nearly annihilated in the Strait of Tsushima. It was all over in less than an hour. Of a combined Russian force of twenty-nine ships, only two destroyers and a light cruiser managed to escape. The Japanese lost a total of three torpedo-boats, and had only 117 casualties, compared to the Russians' 4830. Seen here is a Japanese cruiser. That a small Asiatic people had crushed an arrogant, white colossus, was a shock to many.

Woman cyclist in pants around 1900.

Two ballerinas receiving attention backstage.

A rough and tough tumbler from around 1900.

144. SOUTHEASTERN EUROPE AFTER THE BALKAN WARS 1912-13
—— Ottoman Empire's western boundary in 1912

The Balkan Wars 1912–13 and the Assassination in Sarajevo 1914

October 8, 1912 Montenegro declared war on Turkey. Within a week, Serbia, Bulgaria and Greece had been drawn into the war. By December 3 a cease-fire had already been reached, after the Balkan states had been victorious on all fronts. At the peace of London on May 30, 1913, Turkey was forced to give up all her European territories, with the exception of those nearest Constantinople. Albania became an independent state.

Bulgaria was dissatisfied with her acquisitions, and only a month after the peace went to war against a Serbian and Greek alliance. Romania joined on the side of Greece and Serbia. Turkey attacked Bulgaria as well.

In August 1913 a peace agreement was reached in Bucharest. Turkey broadened her European bridgehead, Romania received a smaller Bulgarian territory on the Black Sea, while Greece and Serbia divided the greater part of Macedonia between them (map nr. 144). Austria was worried that a stronger Serbia had become a focal point for Slavic nationalism.

In June of the next year the Serbian ambassador in Vienna warned the Austrian government against a visit by Arch-Duke Franz Ferdinand to Bosnia. He suspected that Serbian nationalists had planned an assassination. The warning was ignored, and on June 28 the Austrian crown prince and his wife were shot by a Bosnian student, Gavrilo Princip, in the capital of Sarajevo (illus. at left), triggering the First World War.

The Near East 1907–14

1907 After years of bitter struggle between Russia and Great Britain in **Persia (Iran)** the country is partitioned into spheres of influence (map nr. 145). Russia also accepts Brit. hegemony in Afghanistan, with India overseeing its foreign policy.

1908 In a coup d'etat in the **Turkish (Ottoman) Empire** students and officers («young Turks») depose the sultan.

1911–12 In the war with the Turks, the Italians occupy **Libya**, which after the peace of Ouchy at Lausanne (1912) becomes an Italian colony.

1912 Egypt, which has been occupied by the British since 1882, becomes a British protectorate.

145. MIDDLE EAST 1914
▢ The Ottoman Empire
▢ British ▢ French

The Wright brothers fly for 59 seconds in North Carolina. 17 December. 1903.

British warplane from the First World War.

A triple-decker from 1918.

146. EUROPE IN THE GREAT WAR 1914-18

The Central Powers

Neutrals that joined the Central Powers during the course of the war, with date

The Entente (Allied) Powers

Neutrals that joined the Allies during the course of the war, with date

Neutral throughout the war

Sea areas where Germans declared total U-boat war from 1917

Central Power's offensives

Extent of offensive with date:

Eastern front   West. front   Southern Europe

Allied offensives

Extent of offensive with date

The First World War

A month after the assassination of Franz Ferdinand in Sarajevo on June 28, 1914 (see p. 103) Austria-Hungary declared war on Serbia. The map above shows at which points and on which side the other European countries became involved.

The war was the first conflict in history to involve the entire world (map nr. 149). The war was fought in several countries, on all the oceans, and, for the first time, aircraft were used extensively. **The water colour** by Marcel Jeanjean

at the top of the following page shows a training camp in France. After a lightning offensive by the Germans in the summer and autumn of 1914 (map nr. 147), the troops dug in, in trenches on both the Western and Eastern fronts. A portion of Paul Nash's **painting at left** depicts conditions in a British trench on the Western Front. They gradually became fortified field positions, with bomb shelters, connecting passageways, etc., and with complicated technical equipment like periscopes, listening devices, and the like.

But large battles were also fought,

both on land and at sea. The first of these on the Western Front, which took place September 5–12, 1914 on the Marne between German and combined French and British forces, halted the German offensive and saved Paris. In the first battle on the Eastern Front, at Tannenberg, August 30, 1914, the Germans handed the Russians a decisive defeat. **The caricature above** shows German crown prince Frederick Wilhelm and his father, Emperor Wilhelm II, peering toward Verdun from atop a pile of corpses, and agreeing that they will need more of these before they are

British airship from
the First World War.

German soldier with
a hand grenade.
1917.

British tank from
the First World War.

147. THE EASTERN FRONT 1914-17

```
---    Extent of Russian offens. 1914
Extent of Cent. Power's
offens. ......  April -15
       June -15 ---- Sept. -15
—      Front Sept. -16
—      Front Dec. -17
       (armistice)
```

The Human Cost

Country	Mobilized	Fallen
Germany	11 mill.	1,8 mill.
Russia	12 mill	1,7 mill
France	8,4 mill.	1,3 mill
Austria-Hungary	7,8 mill.	1,2 mill.
Italy	4,6 mill.	500 000
USA	4,4 mill.	125 000
Total for these countries	48,2 mill.	6,6 mill.

In all 8,5 million people lost their lives in the First World war.

148. WESTERN FRONT 1914-18

```
⟹    German offensive 1914
·····  German position Sept. 1914
—     Front line 1914-15
▲▲▲   Front line Dec. 1917 (Siegfried
– – –  Front line Oct. 1918    Line)
—     Front line Nov. 1918
```

high enough to see the city. The battle of Verdun lasted from February 21 until December 16, 1916, and cost the French ca. 380 000, and the Germans nearly 340 000, killed, wounded and missing.

At eleven o'clock on November 11, 1918 a armistice went into effect. The Central Powers had been defeated. **Map nr. 151, p. 107** shows some of the consequences of the Treaty of Versailles in 1919.

149. ALLIANCES IN THE GREAT WAR 1914-18

☐ Central Powers and allies	☐ Allies at war's end
☐ Allied Powers Jan. 1917	☐ Neutral states

French field ambulance from the First World War.

A revolutionary soldier speaking to workers in Petrograd in July 1917.

Members of the women's Bolshevik «Death Battalion» in 1917.

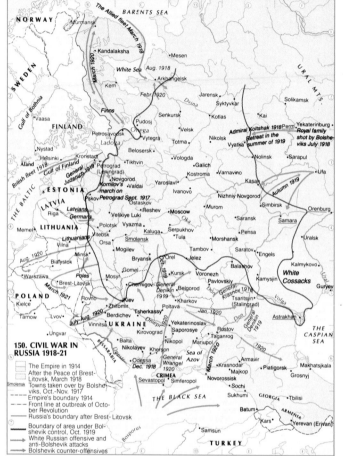

NORWAY
BARENTS SEA
Murmansk
The Allied fleet March 1918
Kandalaksha
Mesen
White Sea Aug. 1918
Kem
Arkhangelsk
Febr 1920
SWEDEN
Jarensk
Solikamsk
Dvina
URAL MTS.
Helsinki
Senkursk Kotlas
Kai
Syktyvkar
Vaasa
Gulf of Bothnia
FINLAND
Pudosj
Velsk
Nikolsk
Perm Yekaterinburg
Admiral Koltshak 1918
Petrosavodsk
Onega
Vytegra
Totma
Vyatka
Retreat in the
summer of 1919
Royal family
shot by Bolshe-
viks July 1918
Nystad
Kronstadt
Ladoga
Belosersk
Vologda
Galich
Nolinsk
Sarapul
Aland
Gulf of Finland
General
Judenitsj 1919
Tikhvin
Novgorod
Valdai
Yaroslavl
Kostroma
Varnavino
Kasan
Ufa
British fleet 1918
ESTONIA
Petrograd
(Leningrad)
Pskov
Kornilov's
march on
Petrograd Sept. 1917
Ostaskiv
Ivanovo
Nizhniy Novgorod
Autumn 1919
Orenburg
LATVIA
Riga
Latvians
Germans
Reshev
Moscow
Murom
Simbirsk
Samara
THE BALTIC
Velikiye Luki
Oka
Saransk
Memel
LITHUANIA
Polotsk
Vyazma
Kaluga
Serpukhov
Pensa
Uralsk
Lithuanians
Vitebsk
Smolensk
Tula
Morshansk
Vilna
Orsa
Tambov
Saratov
Engels
Minsk
Mogilev
Bryansk
Orel
Jelez
Balashov
Kalmykovo
Bialystok
Gomel
Kursk
Voronezh
Kamysjin
White
Cossacks
Warszawa
Poles
Mosyr
Chernigov
Belgorod
General
Denikin
1919
Pavlovskiy
General
Krasnov 1918
Tsaritsyn
(Stalingrad)
Guryev
Brest-Litovsk
POLAND
Rovno
Zhitomir
Kiev
Tsherkassy
Kharkov
Poltava
Volga
Astrakhan
Kielce
March 1921
Berdichev
Dniepr
Jan. 1920
Don
General
Denikin
THE
CASPIAN
SEA
Tarnow
Lvov
July-Aug. 1920
Vinnitsa
UKRAINE
Yekaterinoslav
Saporosje
Rostov
Ungvar
Kirovograd
Taganrog
BESSARABIA
Balta
Nikolayev
Kherson
Nikopol
Mariupol
Sea of
Azov
150. CIVIL WAR IN
RUSSIA 1918-21
Odessa
General
Wrangel
1920
Krasnodar
Armavir
Piatigorsk
Makhatsjkala
The Empire in 1914
After the Peace of Brest-
Litovsk, March 1918
Towns taken over by Bolshe-
viks, Oct.-Nov. 1917
Empire's boundary 1914
Front line at outbreak of Octo-
ber Revolution
Russia's boundary after Brest- Litovsk
Boundary of area under Bol-
shevik control, Oct. 1919
White Russian offensive and
anti-Bolshevik attacks
Bolshevik counter-offensives
Dec. 1918
CRIMEA
Sevastopol
Simferopol
Majkop
Novorossiisk
Sochi
Grosnyj
THE BLACK SEA
Sukhumi
GEORGIA
Tbilisi
Batum
ARMENIA
Yerevan (Erivan)
Bosporus
Samsun
Kars
TURKEY

Above: Bolshevik leader Vladimir Lenin, and Joseph Stalin, in Petrograd, April 16, 1917.

Civil War in Russia 1918–21

After famine had led to the Feb. Revolution in March 1917, and when Nicholas II abdicated, there was chaos in the old Russian empire, leading to civil war in 1918. The opponents of the Bolsheviks formed an anti-Bolshevik government in 1918 in Samara on the Volga. In September this provisional government joined another, strongly conservative one in Omsk, under admiral Alexander Koltshak.

In the south, general Anton Denikin had assembled an anti-Bolshevik army, and south of the Gulf of Finland the «White» forces under general Nicolai Yudenitch were standing ready. The White Army made great progress in 1919, and Denikin's successor, general Piotr Wrangel advanced from the south again in January 1920. But by now the Bolshevik counter-offensives were in full swing. By August 1920 the Red Army was victorious.

Right: The Russian imperial family, photographed in the park at the summer palace of Tzarskoye Selo, southeast of Petrograd, in January 1916. From left: the tzar's adjutant, Tzar Nicholas II, followed by his daughters, Grand Dutchess Tatyana, Olga, Maria and the lively Anastasia. Farthest right, in naval uniform, Crown Prince Alexei (b. 1904). The boy in the foreground, and the three in the background, are sons of the Tzar's eldest sister. In July 1918 Nicholas, his wife, his four daughters and Prince Alexei were shot by the Bolsheviks at Yekaterinburg.

Italian Fiat from 1936.

Woman's fashion – «La Garçonne» style – from end of the 1920s.

General Foch leads the sisters Alsace and Lorraine back to France.

Above: Croydon outside London. Painting by Kenneth McDonough from the 1920s. There was rapid technical development during WW1 and passenger traffic grew just as rapidly after the war. In 1919 the first London–Paris route was opened.

The Peace of Versailles 1919

Germany, held chiefly responsible for the war, lost all colonies, Alsace-Lorraine, and Northern Schleswig (Dan. 1920). Saarland under the control of the League of Nations forced to cede large territories to the newly reconstituted Polish state (map nr. 151). Danzig a free state. Memel region to Lithuania. Austria-Hungary was completely dissolved (see map).

Italian Imperialism

1889–90 Italy acquires Somaliland (1889) and Eritrea (1890) (map nr. 152).

1896 Italian troops defeated at Adua. Ethiopia, then an Italian protectorate, gains independence.

1911–12 Libya and the Dodecanese Is. (incl. Rhodes) are conquered.

1919–24 Tyrol, Istria, Zara and Lagosta are annexed.

1936 Ethiopia is occupied, emperor Haile Selassie flees to Great Britain.

1939 Albania is occupied.

1941 The Italians in Ethiopia surrender to the British. Haille Selassie returns home.

151. CENTRAL AND EASTERN EUROPE AFTER THE GREAT WAR *(From 1925.)*

1914 boundaries of:
— Germ. — Russia
— Austro-Hungary
-- Curzon Line
Year of accession given

152. MUSSOLINI'S EMPIRE

Colonies 1880's - 90's
Protectorate 1889-96
Conquests 1912-24
Mussolini's conquests 1936-41

Radio with speaker from around 1925.

Adolf Hitler chose the swastika as his magic symbol.

Distribution of food to the unemployed in the period between the wars.

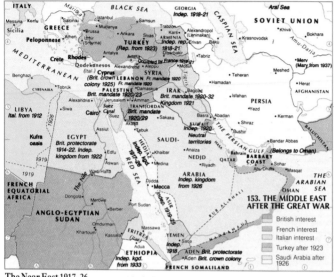

153. THE MIDDLE EAST AFTER THE GREAT WAR

- British interest
- French interest
- Italian interest
- Turkey after 1923
- Saudi Arabia after 1926

Germany 1918–39

Following defeat in the First World War the radical left hopes for a development along Soviet lines, but elections in January 1919 confirm that a majority of Germans are in favour of a non-socialist democratic republic. A new constitution is adopted at Weimar. Friedrich (Fritz) Ebert becomes the first president of the Weimar Republic, and Philip Scheidemann becomes chancellor. The harsh penalties imposed by the Treaty of Versailles have created a desire for revenge in the country, and Scheidemann leaves office in protest.

1920 Adolf Hitler becomes propaganda chief in a new political party: The National-Socialist German Workers' Party (Nazis).

1923 French and Belgian troops enter the Ruhr region to put pressure on Germany, which is having trouble meeting payments on its colossal war reparations debt. The country is experiencing an absurd rate of inflation. Just before the government puts a stop to it through devaluation in November, the rate of exchange in U.S. dollars is climbing at a rate of 613 000 marks per second on the stock exchange! Adolf Hitler carries out an unsuccessful coup d'etat in Munich and is imprisoned.

1929 Mass unemployment and social unrest provide ideal conditions for the Nazis' rise to power.

The Near East 1917–26

1917 Hidyaz gains independence.
1918 Yemen, Armenia and Georgia gain independence.
1920 Palestine and Trans-Jordan come under British mandate. French mandate over Lebanon and Syria. Kuwait gains independence. **1921** Iraq gains independence. Georgia becomes part of the Soviet Union.
1922 Egypt becomes an independent kingdom.
1923 At the Peace of Lausanne, Armenia, the Izmir region and a portion of the peninsula west of Istanbul are accorded Turkey, which becomes a republic.
1926 Saudi-Arabia becomes an independent kingdom, including Asir and Hidyaz.

Above: «Peace in our time,» prime minister Chamberlain said when he returned to London following his meeting with Adolf Hitler in Munich on September 29, 1938. The Sudetenland was to be sacrificed and annexed by Germany. When Chamberlain announced the declaration of war against Germany on September 3, 1939, he admitted his error in judgment.

154. FORMS OF GOVERNMENT IN EUROPE 1938

- Democracy
- Communist
- Fascist
- Other authoritarian government

Nazi Germany used the old Germanic «Heil!» salute.

From the field hospital, 110 m deep on the French Maginot Line.

1933 Hitler is elected chancellor, and quickly has all other political parties dissolved.

1934 Aging president Paul von Hindenburg dies. Hitler assumes his office as well, calling himself «Führer and Chancellor of the Reich».
Political opponents are sent to concentration camps. Jews are singled out for especially brutal treatment (illus. bottom of page).

1935 In contradiction to the Treaty of Versailles, universal military service is introduced.

1936 In March German troops advance into the demilitarized Rhineland (map nr. 155). In July, civil war breaks out in Spain (map nr. 156). Francisco Franco gets support from the German air force and Italian troops. In October Hitler visits Benito Mussolini in Rome, where they reach a secret agreement on a parallel foreign policy (illus. right).

1938 In March Hitler annexes Austria, and in October the Sudetenland, both moves allowed by Munich Agreement (illus. p. 108).

1939 In March Germany takes the remainder of Czechoslovakia, and in August signs a non-aggression pact with Russia (illus. right). On September 1 German troops invade Poland. Two days later the Second World War is underway.

Bottom: Jews on their way to a concentration camp. From a book from 1935, intended to teach children that Jews were rabble.
Farthest right: A meeting between Hitler and Mussolini. **The caricature** *of Stalin and Hitler carries the caption: «How long will the honeymoon last?»*

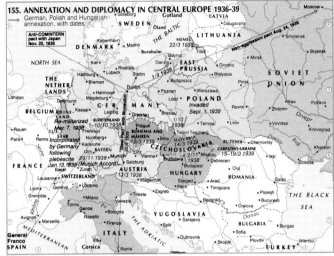

155. ANNEXATION AND DIPLOMACY IN CENTRAL EUROPE 1936-39

German, Polish and Hungarian annexation, with dates

Anti-COMINTERN pact with Japan Nov. 25, 1936

156. THE SPANISH CIVIL WAR

Nationalist control 1936
Conquests up to:
Oct. 1937
July 1938
Febr. 1939

Republican control Feb. 1939
Nationalist stronghold
Republican stronghold
Dates indicate when captured

Nazi stormtrooper.

Mother and child watch nervously for bombers over Madrid.

German Stuka on the attzok in WW2.

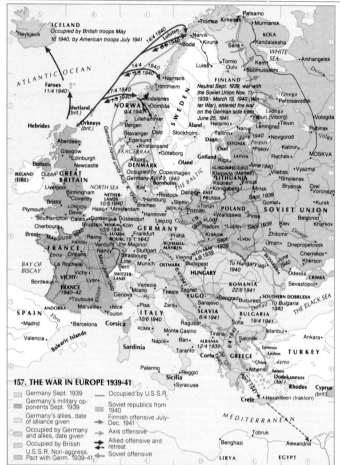

157. THE WAR IN EUROPE 1939–41

- Germany Sept. 1939
- Germany's military opponents Sept. 1939
- Germany's allies, date of alliance given
- Occupied by Germany and allies, date given
- Occupied by British
- U.S.S.R. Non-aggress. Pact with Germ. 1939–41
- Occupied by U.S.S.R.
- Soviet republics from 1940
- Finnish offensive July–Dec. 1941
- Axis offensive
- Allied offensive and retreat
- Soviet offensive

May 10 Germans invade The Netherlands and Belgium. British occupy Iceland.

June 10 Italy enters the war on German side.

June 14 Paris falls.

June 22 France signs an armistice. Most of the country is occupied. Southern France becomes in effect a German dependency.

June–September «Battle of Britain», with massive German air attacks. Huge German losses. Plans for an invasion of England have to be abandoned.

1941: June 22 Hitler attacks the Soviet Union.

Map nr. 157 shows when other European countries entered the war.

Top of next page: In 1939 there were already six concentration camps in Germany with 20 000 political prisoners. **Map nr. 158** shows some of the larger camps during the war, and how many Jews from each were liquidated by the Nazis. It is estimated that over eight million people lost their lives in camps like these.

The picture next to it shows what the British found when they entered the concentration camp at Bergen-Belsen on April 17, 1945.

Bottom of next page: Those responsible for the attack on Pearl Harbor on December 7, 1941. From a caricature in «Fortune». The man with the glasses is Japan's prime minister, general Hideki Tojo, with admiral Chuichi Nagumo. Tojo was judged to bear individual responsibility for the war and hanged in November 1948.

Below, this page: Tower Bridge near the London docks during the blitz of 1940. From a painting by Charles Pears.

Europe 1939–41

1939 September 1 German troops invade Poland from the west and from East Prussia.

September 3 Great Britain and France declare war on Germany.

September 17 The Soviet Union occupies the eastern part of Poland.

November 30 The Soviet Union attacks Finland.

1940 March 12 The Finns capitulate. The «Winter War» comes to an end.

April 9 Denmark and Norway attacked.

110 THE SECOND WORLD WAR 1939–41

An American carrier in the Pacific Fleet.

6 August 1945, the first atombomb is dropped over Hiroshima.

American soldiers planting their flag on a Pacific island in 1945.

158. MAJOR CONCENTRATION CAMPS IN GREATER GERMANY 1939-45

NORWAY 750
SWEDEN
DENMARK 100
LATVIA 70 000
LITHUANIA 104 000
SOVIET UNION
THE NETHER-LANDS 104 000
Neuengamme
Ravensbrück
Stutthof
POLAND 750 000
BELGIUM 40 000
Bergen-Belsen
GERMANY
Mittelbau
Sachsenhausen
Treblinka
Sobibor
Buchenwald
Chelmno
Gross-Rosen
Maidanek
Natzweiler
Theresienstadt
Flossenburg
Belzec
Auschwitz
FRANCE 65 000
Dachau
Mauthausen
CZECHOSLOVAKIA 60 000
SWITZER-LAND
ITALY 9 000
AUSTRIA 50 000
HUNGARY 700 000
ROMANIA 500 000
YUGOSLAVIA 58 000
BULGARIA

Great Germany
Figures: Nr. who perished

The Far East 1941–45

1941 December 7 Japanese co-ordinate attacks on the American naval base at Pearl Harbor, Hawaii, and bases in the Philippines.

December 8 The U.S. and Great Britain declare war on Japan. On **December 11,** both Germany and Italy declare war on the USA.

1942 The Philippines, Malaya and Singapore are taken, and parts of Burma and Thailand are occupied by Japan. A number of islands in the Pacific and the Bismarck Archipelago are also taken (map nr. 159).

June 3–6 Allied victory in Battle of Midway a turning point. Allies on the offensive under general Douglas MacArthur. One group of islands after the other is retaken (map nr. 160).

1944: October 23–27 The Allies defeat the Japanese fleet at Leyte.

1945: August 6–9 U.S. drops A-bombs over Hiroshima and Nagasaki.

August 15 Japan capitulates.

159. THE SECOND WORLD WAR IN THE FAR EAST, DEC. 1941-AUG. 1942

Japanese control Dec. -41
Allied control Dec. 1941
Limit of Japanese power Aug. 1942
Japanese base and offensive
Allied base and offensive
Sea battle won by Japanese
Sea battle won by allies

160. THE WAR IN THE FAR EAST, DEC. 1941-AUG. 1942

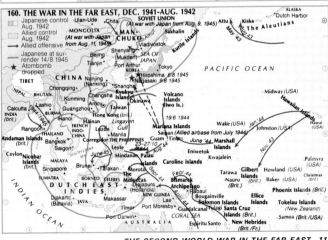

Japanese control Aug. 1942
Allied control Aug. 1942
Allied offensive from Aug. 11, 1945
Japanese at surrender 14/8 1945
Atombomb dropped

German paratroopers over The Netherlands, 10 May 1940.

 A British «desert rat» in North Africa.

 A German V2 rocket, ready for launching.

161. THE WAR IN EUROPE 1942-45

- The Axis Powers 1942
- Joined the Allies, with date
- German control 1942
- Allied control Nov. 1942
- Neutrals that joined Allies
- Neutral throughout war
- — Allied offensive
- Front line: — Oct. 1943
- — Dec.-44 — March -45

Kirkenes • Petsamo
Narvik • Murmansk
Kola Peninsula
Kiruna
Salla •
Rovaniemi •
WHITE SEA • Arkhangelsk
Trondheim
FINLAND
Armistice with the Soviet Union Sept. 4, 1944. At war with Germany Sept. 9, 1944 - April 25, 1945
Petrosavodsk
Dvina
Onega
Shetlands
NORWAY
• Bergen
Oslo •
SWEDEN
Helsinki
Ladoga
SOVIET
Stavanger •
Stockholm
Tallinn
Dagö ESTONIA
Leningrad
Schlüsselburg
• Kasan
Glasgow
Göteborg
Ösel
Riga
LATVIA
Peipus
Novgorod
• Gorki
• Edinburgh
NORTH SEA
DENMARK
Malmö
Velikiye Luki
Pskov
Demjansk
Moscow
GREAT BRITAIN
Copenhagen
THE BALTIC
Memel
LITHUANIA
Vitebsk
Rsjev
Vyazma
Oka
UNION
London •
THE NETHER-LANDS
Hamburg
Königsberg
Kaunas
Smolensk
Mogilev
Orel
Voronezh
Amsterdam
Danzig
Goldap
Vilnius
• Southampton
Arnhem
Stettin
Berlin
GER-MANY
Bialystok
Minsk
Gomel
Kursk
Antwerpen
BELGIUM
Elbe
Warszawa
Brest-Litovsk
Belgorod
Caen Dieppe
Brussels
LUXEM-BOURG
Torgau
Dresden Breslau
Kiev
Kharkov
Stalingrad Nov. 1942-febr. 1943
Reims
Frankfurt
Wisła
Lvov
Tarnopol
Bug
Paris
Nürnberg
Prague
BOHEMIA-MÄHREN
Nikopol •
Saporosye
Rostov
Loire
FRANCE
Strasbourg
München
Linz
Vienna
SLOVAKIA
Dnestr
Dnepr
Vichy
Bern
SWITZER-LAND
Budapest
HUNGARY 31/12 1944
Iaşi •
SEA OF AZOV
Krasnodar
Oradour
Lyon
Milano
Trieste
Zagreb
Debrecen
ROMANIA 25/8 1944
Odessa
Crimea
Kerch
Rhône
Po
Bologna
Jajce
CROATIA
Beograd
Bucureşti
Sevastopol
Jalta
Marseille
La Spezia
Ravenna
SERBIA
Ploieşti
Donau
THE BLACK SEA
SPAIN
Toulon
Pisa
ITALY 13/10 1943
MONTE NEGRO
Sarajevo
BULGARIA 8/9 1944
Barcelona
Corsica
Roma
Anzio
Nettuno
Napoli
Armistice Sept. 8, 1943
ALBANIA
Tirana
Sofia
İstanbul
Ankara 23-2 1945
Balearic Islands
Sardinia
Taranto
Saloniki (Thessaloníke)
GREECE
TURKEY
American and British troops
Bizerte
Sicilia
Messina
Reggio
Catania
Athens
Alexandrette
Haleb (Aleppo)
SYRIA
ALGERIA
Tunis
British troops July 1943
Gela
Siracusa (Syracuse)
Malta (Brit.)
British troops Oct. 1944
Rhodes (Ital.)
Cyprus (Brit.)
Beirut
LEBANON
Damascus
TUNISIA
Gabès
MEDITERRANEAN
Crete
PALESTINE
Jerusalem
TRANSJORDAN
• Tripoli Jan. 1943
Benghazi Nov. 1942
CYRENAICA
Tobruk
Alexandria
Port Said
SAUDI-ARABIA 1/3 1945
LIBYA
El Alamein Okt. 1942
EGYPT
Cairo
Suez

Above: In the winter of 1942–43 Russian and German troops battled for Stalingrad. The Russians were stubborn, and on January 31 German general Friedrich von Paulus and 90 000 men had to surrender. Walter Molina's drawing shows a German tank in Stalingrad.
Below: Bernard Law Montgomery, who with his 8th Army defeated the Germans at El Alamein in 1942, continued through North Africa, on to Sicily in July 1943, and then to southern Italy in September (maps nrs. 161 and 162). From a painting by Sir Oswald Birley.

4/11 1942 Ⓐ Algiers
Bougie
12/11 1942 Ⓑ
Bône
Bizerte
7/5 1943
Tunis
Ⓒ Messina
Sicily
Reggio
Syracuse
Ⓓ Athens
GREECE
TURKEY
Cyprus Ⓘ (Brit.)
Biskra
TUNISIA
Kasserine 18/2 1943
Gabès
Malta (Brit.)
Crete
ALGERIA
Medenine 16/2 1943
Tripoli
MEDITERRANEAN
Port Said
Tarhuna 19/1 1943
Benghazi 19/11 1942
CYRENAICA
Tobruk 12/11 1942
Sidi Barrani 9/11 1942
Alexandria
Suez
Sirte 19/11 1942

162. THE ALLIES' CAMPAIGNS NOV. 1942-MAY 1943 OPERATION TORCH

- Allied advance, date of Axis surrender
- British forces in Operation Torch

21/12 1942
El Agheila 16/12 1942
LIBYA
El Alamein 2/11-5/11 1942
EGYPT
Cairo

American infantry-
man with gas mask.

American bomber – a
Boeing B-17.

American landing-
craft in Normandy.
6 June 1944.

163. THE EASTERN FRONT 1942-45
Greater Germany
Germany's allies
German control Nov. 1942
Front line:
Spring 1943   Autumn -44
Autumn 1943   May 1945

Western and Eastern Europe 1942–45

1942

November 2–5 General Rommel's German troops are defeated by the British at El Alamein.

November 8 Allies land in Morocco and continue eastward.

November 11 The remainder of France occupied by Germany.

1943

January 31 Germans capitulate at Stalingrad. **July 10** British land in Sicily (map nr. 161).

July 25 Mussolini is deposed and replaced by Pietro Badoglio.

September 3 Allied invasion in southern Italy. Cease-fire from September 8.

September 12 Mussolini is freed by German paratroopers and becomes head of a northern Italian fascist republic.

October 13 Italy declares war on Germany. The country is occupied by German troops.

1944

March 19 Germany occupies Hungary.

June 4 The allies take Rome.

June 6 «D-day». Allies invade Normandy (map nr. 165).

July 20 Rebel German officers carry out an unsuccessful assassination attempt against Hitler.

August 24 Paris is liberated.

August 25 Romania declares war on Germany.

September 4–19 Armistice between Finland and the Soviet Union.

September 8 Bulgaria declares war on Germany.

September 11 The Western allies invade Germany.

October 4 British invade Greece.

October 19 Soviet troops enter East Prussia.

1945

February 11 The Russians take Budapest.

March 7 American forces cross the Rhine.

April 7 Russians take Vienna.

April 29 The Germans capitulate in Italy.

April 30 Hitler commits suicide.

May 7 Unconditional German capitulation.

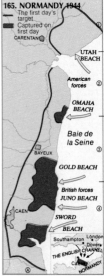

165. NORMANDY 1944
The first day's target
Captured on first day

CARENTAN

UTAH BEACH

American forces

OMAHA BEACH

Baie de la Seine

BAYEUX

GOLD BEACH

British forces

JUNO BEACH

SWORD BEACH

CAEN

Southampton   London
Dover
THE ENGLISH CHANNEL
NORMANDY

164. POWER GROUPINGS IN THE SECOND WORLD WAR, BEFORE AND AFTER PEARL HARBOR

Germany, Italy, Japan and allies:
- bef. Dec. 7, 1941
- after Dec. 7, 1941

The Allies:
- bef. Dec. 7, 1941
- after Dec. 7, 1941

Date of joining Allies given   Neutral throughout war

A. Albania
B. Belgium
Bu. Bulgaria
D. Denmark
H. Hungary
L. Lebanon
N. The Netherlands
S. Switzerland
Sl. Slovakia
Y. Yugoslavia

British medical team carrying a casualty to safety.

Soviet citizens, hanged by the Germans before their withdrawal.

Emaciated prisoners in a German concentration camp.

166. GERMANY AFTER THE SECOND WORLD WAR

Zones of occupation 1945-54
British
American
French
Russian
Zones of admin. after 1945
Germany's boundaries 1937
Oder/Neisse Line
Boundary East - West Germ.
Land boundary W. Germ.
district boundary in East
Capital of Land/district

The Human Cost of WW2

It is estimated that nearly 40 mill. perished in WW2, i.e., over four times as many as in World War 1. And this time many more civilians had been killed: approximately half of all casualties in Europe. Ca. 8 million of these died in concentration camps (see p. 110). Estimated war-dead in some of the countries involved:

The Soviet Union	ca. 20	mill.
Germany	ca. 6	mill.
Poland	ca. 5	mill.
Japan	ca. 2	mill.
Yugoslavia	ca. 1,7	mill.
France	ca. 0,6	mill.
Romania	ca. 0,6	mill.
Italy	ca. 0,4	mill.
Great Britain	ca. 0,4	mill.
USA	ca. 0,3	mill.

Right: A 15-year-old in a Luftwaffe uniform sobs after hearing the announcement of the German surrender in May 1945. When military reserve strength in the final months of the war had become practically nonexistent, young boys – not least members of the youth corps «Hitlerjugend» – were drafted tb defend the Fatherland.
Below: Three representatives of the allies in Berlin – a British private, a Russian woman from the Red Army and an American corporal – offer one another their hands in joy upon hearing the report of Japan's capitulation on August 15, 1945. The war had finally come to an end.

Peace Negotiations and the Division of Germany

At war's end it had already been decided that Poland was to get the eastern half of the old Germany as far west as the Oder-Neisse (map nr. 166). East Prussia was to be divided between Poland and the Soviet Union. The rest of Germany was to be divided into four occupied zones: one American, one British, one French and one Russian. From 1949 on the Soviet zone was an independent state under the name Deutsche Demokratische Republik (East Germany). In the same year the three other zones became the state of the Bundesrepublik Deutschland (West Germany).

AIRLIFT 28/6 1948 - 12/5 1949. The Allieds answer on Sovjet blockade of the supply-lines.

167. BERLIN AFTER THE SECOND WORLD WAR
The Berlin Wall with checkpoints
SPANDAU Districts (Bezirke)
Railway   Major road

KEY TO NUMERALS:
① Brandenburg Gate
② Checkpoint Charlie
③ Free University of Berlin
④ Allied Administrative Council
⑤ Humboldt University

168. ECONOMIC ALLIANCES IN EUROPE AFTER 1945

Members of COMECON, Council for Mutual Economic Assistance, founded Jan. 1949. Year and map text indicate later member country, associate country or ex-member country.

Members from outside Europe: MONGOLIA (joined 1962), CUBA (joined 1972) and VIETNAM (joined 1978).

Members of OECD, Organization for Economic Co-operation and Development, founded Sept. 1961 by reorganization of OEEC founded 1948.

Members from outside Europe: AUSTRALIA, JAPAN, CANADA, N. ZEALAND and USA.

Members of Council of Europe, founded May 1949.

EFTA Members of EFTA, European Free Trade Association, founded May 1960.

EEC Members of EC, The European Community, founded July 1967 through union of Coal and Steel Union (1952), European Economic Community (1958) and EURATOM, European Atomic Energy Commission (1958)

Members of Benelux, founded Jan. 1948.

CELAND Members of The Nordic Council, founded 1952. Finland joined in 1955.

Berlin, which lay in the Soviet zone, was also to be divided among the four occupying nations. The «Cold War» between East and West, however, led to Russian construction of a wall right through the city in 1961 (map nr. 167). The wall did not come down again until 1989.

Japan. Just after the capitulation in August 1945, American forces occupied Japan until April 28, 1952. The Japanese had agreed to revoke all claims to territories outside Japan proper.

The United Nations (UN)

Even in the early stages of the war the allies had agreed that the peace to follow would best be preserved through the establishment of an international organization of all the world's free nations.

The idea was discussed during the war, and on April 25, 1945 representatives from fifty countries assembled in San Francisco to outline rules for the formation of the United Nations. Two months later they had reached agreement, and October 24, 1945 (UN-day) the organization was formally inaugurated.

The UN has six main bodies, of which the General Assembly and Security Council are the most important. The organization has its headquarters in New York, and **the picture below** shows the UN Building, built there in 1952.

De Gaulle's France

General Charles de Gaulle (1890–1970) managed to get out of the country when France capitulated in June 1940, and following the liberation in June 1944, he returned to France as head of the provisional government. When, however, his suggestion to give the president extended powers in a new constitution met with stiff opposition, he left the government in 1946.

When in 1958 France found herself on the brink of civil war and chaos, he again took over responsibility for the administration of the country. France's and Europe's strong man for more than a decade (**picture below**).

 The Soviet space-craft «Vostok» from 1962.

 The U-2 incident in 1960 was seen as a collision with the dove of peace.

 Stalin died 2 March 1953 and was laid in «lit de parade» in the Kremlin.

169. MILITARY ALLIANCES IN EUROPE AFTER 1945

- Member of NATO, North Atlantic Treaty Organiz., founded 1949. Incl. of USA and CANADA.
- Member of Warsaw Pact, founded 1955.
- WEU Member of Western European Union, founded 1954 after Western Union was dissolved.
- BP Member of Balkan Pact 1953-60.
- CENTO Member of Central Treaty Organiz., founded 1959 after dissolution of Bagdad Pact.
- Neutral State

The Faroes (danish), FINLAND, Shetland, NORWAY (1949), SWEDEN, ATLANTIC, NORTH SEA, GREAT BRITAIN (1949), EIRE, DENMARK (1949), THE BALTIC, WEU-CENTO, THE NETHERLANDS (1949) WEU, EAST GERMANY, POLAND, BELGIUM (1949) WEU, WEST GERMANY (1949) WEU, CZECHOSLOVAKIA, THE SOVIET UNION, CASPIAN SEA, OCEAN, LUX. (1949) WEU, FRANCE (1949) WEU, SWITZERLAND, AUSTRIA, HUNGARY, ROMANIA, BLACK SEA, ITALY (1949) WEU, YUGOSLAVIA BP, BULGARIA, IRAN CENTO, PORTUGAL (1949), SPAIN (Separte military agreement with USA from 1953), Corsica (fr.), Sardinia, ALBANIA (Member of Warsaw Pact 1955-68), GREECE (1952) BP, TURKEY (1952), BP-CENTO, SYRIA, IRAQ (Member of Baghdad Pact 1955-59), MEDITERRANEAN, Sicilia, Crete, Cyprus (Indep. rep. from 1960), Gibraltar (br.), ALGERIA, TUNISIA

Nato and the Warsaw Pact

The Western powers began shortly after the war to make plans for a common defense alliance, and when the U.S. Senate in 1948 gave its approval for the government to enter into an agreement with Western Europe, a draft agreement was ready in January of 1949.

On April 4, 1949 the Atlantic Treaty was signed by the USA, Canada, Great Britain, France, the Benelux countries, Portugal, Denmark, Norway and Iceland. Since then several other nations have joined the alliance (map nr. 169). Each member nation will treat an attack on another member nation as an attack on itself.

A military and intelligence defense organization was formed which came to be known as NATO (the North Atlantic Treaty Organization).

The Eastern Bloc formed a corresponding military alliance, the so-called **Warsaw Pact**. It was signed in Warsaw on May 14, 1955 by the Soviet Union, Albania, Bulgaria, Hungary, Poland, Romania, Czechoslovakia and East Germany. Albania pulled out of the pact in September of 1968 after having broken with the Soviet Union the previous year and orienting itself more in the direction of China politically.

Right: Relative military strength between NATO and the Warsaw Pact at the beginning of the 1980s. Source: «Le Point», Paris, April 18, 1986.

Soviet Union 1945–85

1945–47 The country has become a great-power, but faces massive problems of reconstruction. The cult of personality surrounding Stalin reaches new heights. Poland, East Germany, Czechoslovakia, Hungary, Romania, Yugoslavia and Bulgaria are controlled from Moscow. Ideological conformity has been achieved.

1948 Berlin Crisis. The Russians block all roads leading to West Berlin. The Western powers carry out an airlift (map nr. 167, p. 114). The «Cold War» heats up.

1949 The country acquires nuclear weapons and is considered a super-power. The arms race has begun.

1953 Stalin dies. Collective leadership, with Nikita Khrushchev as the dominant figure.

1955 The Warsaw Pact is formed.

1956 Uprising in Hungary is put down.

1957–62 Soviet Union gets headstart in the Space-race. Break with China. Crises between the super-powers: Berlin (1958-59), U-2 (1960), Cuba (1962).

1964 Leonid Breshnev takes over.

1968 Intervention in Czechoslovakia.

1970–75 Detente with the West.

1977–80 Somewhat cooler relationship with the USA. The Soviet Union intervenes in Afghanistan (1979). Crisis in Poland following Lech Walesa's founding of the labour union Solidarity (1980).

1982–84 Breshnev dies in November 1982 and is succeeded by Andropov, and after him Chernenko.

1985 Michail Gorbachev is installed as General Secretary of the Communist Party and, in reality, leader of the Soviet Union.

Above: The Anniversary of the Revolution – November 7 – has for many decades been celebrated with military parades in Red Square.
Right: Defense minister Dmitri Ustinov and Leonid Breshnev, during the military parade on Revolution Day 1979.

China 1934–80

1934–35 The first communist state in China – formed in Jiangxi with Mao Tse-tung as foreman in 1931 – is surrounded by forces of the nationalist government, i.e., Chiang Kai-shek's troops. In October 100 000 members of the Red Army succeed in breaking out. Led by Mao, they begin the «Long March» to Shaanxi, which from 1935 becomes the communists' headquarters (map nr. 170).

1937 Japan attacks China. Chiang Kai-shek and the communists make

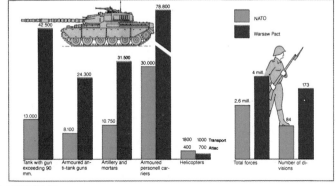

NATO

Warsaw Pact

	Tank with gun exceeding 90 mm.	Armoured anti-tank guns	Artillery and mortars	Armoured personell carriers	Helicopters		Total forces	Number of divisions
NATO	13.000	8.100	10.750	30.000	1800 Transport / 400 Attac		2,6 mill.	84
Warsaw Pact	42.500	24.300	31.500	78.800	1000 Transport / 700 Attac		4 mill.	173

Chinese women in a rice paddy.

Young Chinese girls receiving military training.

After Mao, Chinese young people began dancing to Western rhythms.

Above: In 1966, 72-year-old Mao was venerated as an almost divine figure. As the Cultural Revolution did away with the last vestiges of old-fashioned «bourgeois ideology», it also turned most everything else upside-down. Even down to the traffic lights, where a green light now meant «stop»!

Left: In three years, 350 million copies of chairman Mao's sayings from 1964, the «Little Red Book», had been printed, and became the basis for every aspect of Chinese life.

civil peace to stand united against the aggressor (map nr. 142, p. 102).

1940–45 During WW2 the Chinese receive allied support against Japan (maps 159 and 160, p. 111).

1945 Mao Tse-tung goes on the attack against Chiang Kai-shek and his regime.

1949 The People's Republic of China, with Mao as leader, is proclaimed on October 1, Chiang and his forces withdraw to Taiwan, where in December he establishes his own state – Nationalist China (map nr. 170).

1958 The republic is organized in people's communes. At the same time the «Great Leap Forward» is initiated, an attempt to dramatically increase production.

1963 Conflict with Soviet Union.

1966–70 The «Cultural Revolution» uncovers deeply rooted differences in the party leadership and leads to bloody clashes.

1976 Mao Tse-tung dies and is succeeded by Hua Kuo-feng.

1980 Hua meets severe criticism and is replaced by Hu Yaobang.

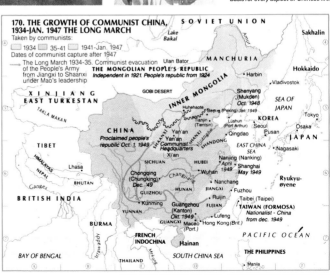

170. THE GROWTH OF COMMUNIST CHINA, 1934-JAN. 1947 THE LONG MARCH
Taken by communists:
1934 ☐ 35-41 ☐ 1941-Jan. 1947
Dates of communist capture after 1947
The Long March 1934-35. Communist evacuation of the People's Army from Jiangxi to Shaanxi under Mao's leadership

A camel carries UNICEF aid to the Gaza Strip.

A child is helped to safety during the civil war in Lebanon.

Lebanese poster urges solidarity with people in Israeli-occupied areas.

171. PALESTINE AFTER THE SECOND WORLD WAR
Division suggested by United Nations, 29/11 1947:
☐ Arab Palestine
▨ Jewish Palestine
▤ Permanent international status
▨ Jewish settlement
☐ Arab states
— Boundary of British Mandate 15/5 1948

172. ISRAEL 1949 ▶
▨ Israel after cease-fire, Jan. 1949
☐ Arab territory, West Bank to Jordan 1950

back, and win large areas of land.
1956-57 Egypt's president Nasser (illus. left) blocks Israeli shipping through the Suez Canal in 1956. The Israelis occupy the entire Sinai Peninsula. Bowing to strong international pressure, Israel pulls out of the Sinai, Gaza and the Strait of Tiran the following year.
1967 Egypt expels the UN troops and closes the Strait of Tiran to Israeli ships. Israel strikes back with great force. In the course of the Six Day War Israel occupies the Sinai Peninsula, Old Jerusalem, the West Bank of the Jordan, and the Golan Heights (map nr. 173).

Right: PLO leader Yassir Arafat (b. 1929), who from 1965 was the leader of the Palestinian resistance organization al-Fatah. Three years later, Arafat was also made leader of the PLO (Palestinian Liberation Organization). Its goal is the establishment of an independent state for the Palestinians. In 1982 the PLO was forced to leave Beirut, but Arafat managed to keep the organization together. **The picture at the top of the page** shows the departure of the PLO's soldiers from Beirut. They dispersed to Tunisia, Syria, the Sudan, Cyprus, Greece and Jordan.

Above: Colonel Gamal Abdel el Nasser (1918–70) was the mastermind behind the coup against Egypt's King Farukh in July 1952. Two years later he became the country's president.

Palestine-Israel 1947–67

1947 A UN-resolution, which has Jewish support, proposes the division of Palestine in a Jewish and an Arab state (map nr. 171).
1948–49 Palestine, which has been under British mandate since 1920, is given up by the British in May 1948. The state of Israel is proclaimed. The Arab countries attack immediately, but the Israelis strike

173. THE SIX DAY'S WAR, JUNE 5-10, 1967 ▶
⦀ UN forces give way May 21
✳ Airfields bombed by Israelis June 5
→ Israeli armoured thrusts
▨ Israel's opponents in Six Day's War
▨ Members of the Arab League
▨ Occup. by Israel in Six Day's War

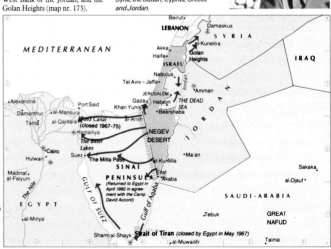

Oil platform in the North Sea.

Armed Iranian woman in traditional Shiite Muslim garb.

World's first atomic powered submarine. The American «Nautilus» (1954).

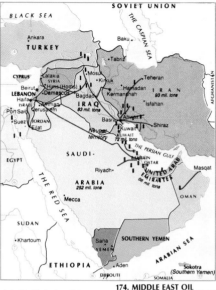

Korea 1945–53

1945 Korea occupied by Russian and American troops (map nr. 175). N. of the 38th Parallel, communist government was established.

1946–47 A UN commission is set to prepare for free elections.

1948 Commission is refused access to the north, but elections are held in the south. Republic of Korea proclaimed on Aug. 15. On Sept. 9 the People's Republic of Korea is proclaimed in the north. On December 12 the UN recognizes the South Korean regime as the country's only legitimate government.

1950–53 North Korean troops attack South Korea (map nr. 175). Armistice in 1953. Korea remains divided.

Left: The Muslim Ayatollah Khomeini, who from exile in Paris led opposition forces in Iran against Shah Mohammed Reza Pahlavi's regime. When the shah fled the country in January 1979, Khomeini returned in triumph to Iran, where he proclaimed the Islamic Republic, with himself as leader. This was the beginning of a period of internal terror and thousands of executions. The following year war broke out with Iraq. There were ferocious tank battles, and many casualties.

174. MIDDLE EAST OIL COUNTRIES

I Important oilfield
• Major oil refinery
— Oil pipeline

Each country's production of crude petroleum in 1986 given in mill. tons

The Oil States of the Mid-East

In **Bahrain** oil was discovered in 1932. **Egypt** has since the 1960s had a steady increase in oil production.

The **United Arab Emirates** have rich oilfields, both on land and offshore. In **Iraq** the first oil deposits were found at Kirkuk in 1927. From here oil pipelines have been stretched to the Mediterranean and the Persian Gulf (map nr. 174). In 1972–73 the internationally owned Iraq Petroleum Co. was nationalized. **Iran** (Persia) nationalized the British oil company Anglo-Iranian Oil Co. in 1951. In **Kuwait** oil production wasn't begun until 1946, but the rich deposits have made the country, which didn't gain independence until 1961, one of the world's wealthiest. In **Oman** the first oil discovery was made in 1964. **Qatar**. British protectorate from 1916 to 1971. Oil discovered in 1939. Production from 1949. **Saudi-Arabia** began extracting oil in 1933, and increased production drastically after 1945.

175. THE KOREAN WAR 1950-53

◀ A. 25/6 - 25/11 1950
→ North Korean offensive July-Aug. -50
— Pusan Bridgehead, 5/8-16/9 -50
→ South Korean and UN offensive Oct.- Nov. 1950
— Limit of greatest advance

▼ B. 26/11 1950 - 27/7 1953
→ Chinese - N. Kor. offens. fr. 16/11 -50
— Southernmost limit in Jan. 1951
→ South Korean and UN counter-offensive, Summer 1951
— Demarcation Line 27/11 1951, confirmed by armistice 27/7 1953

A starving child
from Biafra in
1970.

Slum district in
South Vietnam's
capital of Saigon.

The first human
lands on the moon
in July 1969.

176. RICH AND POOR COUNTRIES AT THE END OF THE 1970'S

Gross National Product per inhabitant (in US $)

- ☐ $ 7000 or more
- ☐ $ 3000-6999
- ☐ $ 700-2999
- ☐ $ 300-699
- ☐ Less than $ 300

Left: Fidel Castro (b. 1927), from 1959 Cuba's prime minister, and from December 1976 the country's president as well. After intensive preparations in Mexico, he came to Cuba in 1956 with 80 men to topple dictator Fulgencio Batista's corrupt regime. The charismatic Castro won the sympathy of the people, and in 1959 he and his guerilla force entered Havana in triumph after Batista had fled. The USA welcomed Castro until he drew closer to the communists (see next page). **Right:** Harvesting sugar cane in Cuba, which is the world's largest producer of sugar.

An American chromeplated gas-guzzler from the 1950s.

A German television set from the early 1950s.

American nuclear power plant in 1979.

Above: *Democrat John Fitzgerald Kennedy (1917–63) represented Massachusetts in the Congress from 1946. Six years later he was elected senator, and in 1960 was nominated as the party's candidate for president. In the election he won a narrow victory over Republican Richard Nixon. Like Castro – with whom Kennedy quickly clashed – the American president too had a strongly charismatic character. John F. Kennedy is remembered, among other things, for the statement: «Ich bin ein Berliner» – I am a Berliner – made during his visit to West Berlin in the summer of 1963. Six months later Kennedy was shot and killed by Lee Harvey Oswald during a visit to Dallas, Texas.*

Cuba 1960–63

1960 Fidel Castro initiates close cooperation with the Soviet Union.

1961 A force of 1 500 Cuban exiles, supported by the USA, go ashore at the Bay of Pigs, but are quickly defeated (map nr. 178). Khrushchev promises Castro support. President Kennedy declares that the USA will counter any Soviet military intervention.

1962 Americans discover Soviet bases for medium distance rockets in Cuba. U.S. naval blockade of Cuba.

1963 Soviet Union gives in. 22 000 Soviet soldiers leave the island. All bases and equipment removed. The Cuban Missile Crisis ends.

177: CENTRAL AMERICA

178: THE CUBA CRISIS, OCTOBER 1962
▲ Russian rocket bases, removed in 1962
⊠ American base
▬ Invasion by Cuban exiles, April 1961

Woman at a farming collective in Nyerere's Tanzania.

Man carrying ammunition on his head during the Biafran War.

Young boy in Angola armed with modern automatic weapons.

A

1939

- EGYPT 1922
- ABESSINIA (Ital. colony) 1936–41
- LIBERIA 1847
- COMMONWEALTH OF SOUTH AFRICA 1910

B

1958

- MOROCCO 1956
- TUNISIA 1956
- LIBYA 1951
- SUDAN 1956
- GUINEA 1958
- GHANA 1957
- ETHIOPIA 1941

C

1965

- ALGERIA 1962
- MAURITANIA 1960
- MALI 1960
- NIGER 1960
- CHAD 1960
- SENEGAL 1960
- CENTRAL AFRICAN REP 1990
- IVORY COAST 1960
- NIGERIA 1960
- CAMEROON 1960
- TOGO 1960
- CONGO STATE 1960
- GABON 1960
- CONGO From 1971 (ZAIRE)
- KENYA 1963
- SOMALIA 1960
- TANZANIA 1964
- ZAMBIA 1964
- MALAWI 1964
- RHODESIA 1965 (Not recognized)
- MADAGASCAR CAR 1960

1. UPPER VOLTA 1960
2. DAHOMEY 1960
3. GABON 1960
4. SIERRA LEONE 1961
5. BURUNDI 1962
6. RWANDA 1962
7. UGANDA 1962

D

1965–90

Canary Islands (Sp.) — Las Palmas

Madeira (Port.)

MOROCCO — Tanger, Ceuta (Sp.), Melilla (Sp.), Rabat, Casablanca, Fez (Sp.), Marrakech

Tunis, Algiers, Oran, Constantine, TUNISIA, Tripoli (Tarabulus)

MALTA 1964, MEDITERRANEAN

Beirut, Tel Aviv, Gaza, Amman, IRAQ, IRAN

WESTERN SAHARA (To 1975 Spanish Sahara. Occupied by Morocco 1975. Under Morocco from 1979. Contested territory.)

ALGERIA — Ain Salah, Tamarasset, Ghat, Murzuk

LIBYA

EGYPT — Alexandria, Cairo, Suez, Aswan

SAUDI-ARABIA, QATAR, UNITED ARAB EMIRATES, OMAN

THE RED SEA

MAURITANIA — Nouakchott

CAPE VERDE 1975 — Praia

GAMBIA 1965, SENEGAL — Dakar

GUINEA-BISSAU 1974 — Conakry

GUINEA — Freetown

SIERRA LEONE, LIBERIA — Monrovia

MALI — Timbuktu, Gao, Bamako

BURKINA FASO — Ouagadougou

IVORY COAST, GHANA, BENIN, Abidjan, Accra

NIGER — Niamey, Kano, Kaduna

Lagos, Ibadan, NIGERIA

CHAD — Ndjamena (Fort Lamy)

Lake Chad

CAMEROON — Yaounde, Douala

CENTRAL AFRICAN REP — Bangui

Omdurman, Khartoum, SUDAN, Kodok (Fashoda)

Massawa, Asmara (To Ethiopia 1952)

Sané, YEMEN, Aden (Brit. to 1967)

Addis Ababa, ETHIOPIA

DJIBOUTI 1977, SOMALIA

Bioko (Macias Nguema) (To Equatorial Guinea 1968), SÃO TOMÉ and PRINCIPE 1975, EQUATORIAL GUINEA 1968, Libreville, Bata, GABON, Brazzaville, Kinshasa (Leopoldville)

Pagalu (Eq. G.)

Biafra (indep. 1967–70), Épinga

ZAIRE (CONGO PEOPLES REP. (RHODESIA) 1964), Kisangani (Stanleyville)

UGANDA, Kampala, RWANDA 1962, BURUNDI 1962, Lake, Tanganyika

KENYA, Nairobi, Mombasa, Dodoma, Dar-es-Salaam

Mogadishu, THE INDIAN OCEAN

Cabinda (District of Angola)

Luanda, ANGOLA 1975, Huambo, Katanga, Lubumbashi, Elisabethville

Seychelles 1976, Zanzibar, Amirante Islands (Brit.), Victoria

TANZANIA

ZAMBIA 1964, Lusaka, Ndola, MALAWI, Lilongwe, Tete, Mozambique

Lake Malawi, COMORO ISLANDS, Aldabra (Brit.)

St. Helena (Brit.)

ZIMBABWE 1980, Harare (Salisbury)

MOZAMBIQUE 1975, MADAGASCAR, Tananarive

Port Louis, MAURITIUS 1968

Ascension (Brit.)

NAMIBIA SOUTHWEST AFRICA 1990, Windhoek, Walvis Bay (S.-A. rep.) (Rep. of S. Africa's administration declared invalid by UN in 1971)

BOTSWANA 1966, Gaborone, Pretoria, Johannesburg

SWAZILAND 1968, Maputo (Lourenço Marques)

LESOTHO 1966, Maseru, Durban

REPUBLIC OF S. AFRICA, Cape Town, Port Elizabeth

ATLANTIC OCEAN

The pictures at the top of this and the next page show, from the left: **1** Archbishop Desmond Mpilo Tutu (b. 1931), who received the Nobel Peace Prize in 1984 for his leadership of the non-violent protest movement against Apartheid. **2** Kenya's colourful leader, Yomo Kenyatta (1891–1978) **3** Julius Nyerere (b. 1922), Tanzania's president 1965–85. He and Kenyatta were among the most outspoken proponents of pan-Africanism and African socialism. **4** Anwar Sadat (1918–81), who became Egypt's leader following president Nasser's death in 1970. **5** Colonel Muammar al-Gaddaffi (b. 1942) came to power in Libya in coup in 1969. Most radical representative of pan-Arabic nationalism. **6** President of the Congo (Zaire), Joseph Kasavubu (1910–69), on the right, and general Sése-Seko (Joseph-Désiré) Mobutu (b. 1930). Kasavubu, who was the Republic of the Congo's first president from 1960 to 1965, was succeeded by Mobutu following a coup d'etat (see Congo Crisis, p. 123).

179. AFRICAN INDEPENDENCE 1939-90

Sovereign state and year of independence

Map A states independent in 1939, map B independent 1939-58 and map C the 25 states that achieved indep. 1959-64. Map D dates show indep. after -64.

- Belgian dependency
- British dependency
- French dependency
- Italian colony
- Portuguese colony, Portuguese province after 1971
- Spanish dependency
- League of Nations mandate, later United Nations

Racial rioting in the Johannesburg suburb of Soweto in 1976.

A political prisoner in Ghana reunited with his wife after five years in prison.

An animal is left dying following a drought in Ethiopia.

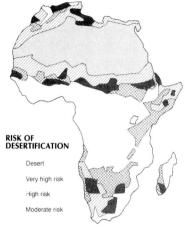

RISK OF DESERTIFICATION

Desert

Very high risk

High risk

Moderate risk

African Liberation 1951–90

1951 Libya (24/12). **1956** Sudan (1/1), Morocco (2/3), Tunisia (20/3). **1957** Ghana (6/3). **1958** Guinea (2/10). **1960 The Great Year of Liberation**: Cameroon (1/1), Togo (27/4), Somalia (26/6), the Congo (Zaire) (30/6), Benin (1/8), Niger (3/8), Upper Volta (Burkina Faso) (5/8), Ivory Coast (7/8), Chad (11/8), Central African Republic (13/8), Republic of the Congo (15/8), Gabon (17/8), Senegal (20/8), Madagascar (21/9), Mali (22/9), Mauritania (28/11), Nigeria (1/10). **1961** Tanzania (1/5) **1962** Burundi (1/7), Rwanda (1/7), Algeria (3/7), Uganda (9/10). **1964** Malawi (6/7), Zambia (24/10), Kenya (12/12). **1965** Gambia (18/2). **1966** Botswana (30/9), Lesotho (4/10). **1968** Swaziland (6/9), Equatorial Guinea (12/10). **1973** Guinea-Bissau (24/9). **1975** Comoro Islands (6/7), São Tomé and Principe (12/7), Moçambique (25/6), Angola (11/11). **1977** Djibouti (27/6). **1980** Zimbabwe (Rhodesia) (18/4). **1990** Namibia (21/3)

The Congo crisis 1960–65

The Congo, formerly the Belgian Congo, became an independent republic on June 30, 1960. The Belgians turned over the government to a coalition, with Patrice Lumumba (illus. below) as prime minister and Joseph Kasavubu as president (illus. above). Conditions quickly became chaotic, however, and Belgian troops entered the country again. In July the mineral-rich province of Katanga (Shaba) seceded under the leadership of Moise Tshombe, as did the southern part of Kasai, led by Albert Kalondshi (map nr. 179). Lumumba asked the UN for help, but when he wasn't satisfied with the efforts of the UN troops, he asked the Soviet Union for assistance. A displeased Kasavubu forced Lumumba out of office in early September 1960.

Shortly afterward, the army, under general Sése-Seko Mobutu seized

Below: The Congo's first prime minister, Patrice Lumumba (1925–61).

power in a coup. But Lumumba's deputy, Antoine Gizenga, formed a strong counter-regime in Stanleyville a few months later. In February 1961 the Tshombe government announced that Lumumba had been found murdered in Katanga. The Soviet Union placed responsibility for the killing on UN secretary Dag Hammarskjöld, who was killed in an unexplained air crash on September 18, 1961 while on his way to a meeting with Tshombe. UN troops were again sent into the country – in all 20 000 men from eighteen countries. In 1964 provinces of Katanga and Kasai gave up their opposition to being incorporated into the Congo.

Order restored when general Mobutu for the second time seized power in 1965. From 1971 on, the Congo is officially known as Zaire.

The map above illustrates the threat of desertification in Africa. Irresponsible logging of the forests, intensive cultivation of the land and overgrazing, combined with almost perpetual drought, turns large areas of land each year into barren wasteland. The same phenomenon represents a threat to certain areas of North and South America, Asia and Australia.

180. THE CONGO CRISIS 1960-65
Areas giving support in 1961 to:

▨ Kasavubu and Mobutu
▨ Albert Kalondshi in Kasai
▨ Lumumba and Gizenga
▨ Moise Tshombe
● UN forces bases 1960-64
Revolts in 1964:
— Limits of Simba revolt
— Limits of Mulélé revolt

♀ Landings by Belgian para-
troops, Nov. 1964
--- Provincial boundary

Vietnamese with child fleeing during the Tet Offensive in 1968.

Farming with ancient methods in South Vietnam.

Woman fleeing war in Laos.

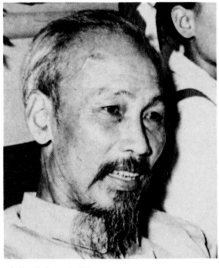

181. INDO-CHINA 1945-54

Members of French Indo-China Union 1949 (Vietnam from –46)

Controlled by Viet Minh:
▬ 1946-50 ▬ 1950-54

☐ After armistice in July 1954

Partition of Vietnam after armistice (17th parallel)

Occupation forces from:
→ Nationalist China
→ United Kingdom
→ France

French Indochina 1945–54

Cambodia

1945 Because Vichy-France (map nr. 143, p. 110) has allowed Japan and Thailand to establish bases in the country during the Second World War, king Norodom Sihanouk, with Japanese approval,
proclaims Cambodia's independence in March. British, Indian and French troops enter the country, and in October they occupy the capital of Phnom Penh (map nr. 181). French rule is established.
1949 France recognizes Cambodia as an independent state within the French Union. The French retain control of the military and foreign policy.
1953 May 9 an agreement is signed with France, insuring Cambodia's full sovereignty in military, judicial, and economic affairs.
1954 The Geneva Convention confirms the country's independence.

Laos

1945 Declared an independent state on April 15. Sisavang Vong is
made king. In August the Chinese invade, occupying most of the country (map nr. 181).
1946 In March French troops enter and retake the country. The government flees to Bangkok. France promises independence.
1947 Laos gets a new constitution. Parliamentary monarchy.
1949 Laos an associate nation in the French Union.
1953 Laos given full independence within the French Union.
1954 Vietminh enter Laos. According to the Geneva Convention, the country is forbidden to form alliances with any country. The government attempts to pacify the pro-communist party, Pathet Lao, which is supported by both the Vietminh and North Vietnam. The French pull out of Laos.

Vietnam

1945 Japanese disarm the French troops and turn over control of the country to a puppet government under emperor Bao Dai. Shortly after the Japanese capitulation, the Vietminh, led by Ho Chi Minh, seize power.
1946 French move toward Hanoi (map nr. 181). An accord is reached, stating that the country will become a free state, but within the French Union. On Nov. 20, exchange of fire between French and Vietnamese soldiers in Haiphong. Three days later a French cruiser bombards a division of Vietnamese soldiers, resulting in a horrendous bloodbath. Vietnamese go on the attack on December 19. The **First Vietnam War** is underway.
1954 Following more than seven years of guerilla warfare – in which from 1950 on France had received considerable support from the USA – the French provoke the Vietnamese into attacking their base at Dien Bien Phu (map nr. 181). After a fifty-five day seige, the French are forced to surrender on May 7. The Geneva Convention negotiates a cease-fire. Vietnam is to be divided at the seventeenth parallel. Ho Chi Minh is elected leader of the Democratic Republic of Vietnam in the north. Bao Dai becomes chief of state in South Vietnam.

Left: When the American war in Vietnam was presented in images like this on television screens in the USA, demands for war-crimes trials were heard, even at home.

An FN soldier moving into Saigon, 30 April 1975. The war is over.

American bomber dropping its load over North Vietnam.

American soldier comforts a North Vietnamese woman.

Left: Vietnamese politician, Ho Chi Minh (1890–1969). After many years in Europe, where, among other things, he had helped to found the French Communist Party in 1920, Ho Chi Minh came to Hong Kong. Here he served as Comintern representative from 1930, and it was here that he laid the groundwork for a communist party in French Indochina (Vietnam). From 1941 he headed the liberation organization Vietminh, and in 1954 he became president of North Vietnam.

Below right: Hanoi is in ruins following the American bombing attack at Christmas 1972.

Vietnam 1955–76

1955 Bao Dai is toppled in South Vietnam. Ngo Dinh Diem is made president. 800 000 North Vietnamese flee south.

1956 Local uprisings in North Vietnam. South Vietnam is supported by USA.

1960 North Vietnamese infiltration of the south. The National Liberation Front (FNL) and its guerilla army are supplied from the north, via the Ho Chi Minh Trail. North Vietnam is supported by China and the Soviet Union.

1964 Purported clashes between American and North Vietnamese naval vessels in the Gulf of Tonkin open the way for increased American involvement. The United States bombs North Vietnam. The total number of bombs dropped is greater than during all of WW II.

1967 450 000 American soldiers are fighting in Vietnam. In this one year, 9 000 of them lose their lives. Massive demonstrations against the war in the U.S. and in other countries.

1968 The Tet Offensive (70 000 men) against the regime in the south is a turning point.

1969 In November president Nixon announces that all American ground troops are to be withdrawn.

1972 North Vietnam carries out a new offensive. U.S. bombing in the north is resumed.

1973 Negotiations in Paris lead to a cease-fire in January, but the armed struggle continues. At this point, three million Americans have taken part in the war, with nearly 54 000 casualties.

1975 South Vietnam's army is in disarray. North Vietnamese and FNL troops take Saigon on April 30. The city is renamed Ho Chi Minh City.

1976 North and South Vietnam are officially reunited on July 2.

B. 1969–73
Armistice, and USA out of Vietnam in Jan. 1973 But fighting continued

182. THE VIETNAM WARS 1954–77

— Partition Line (17th parallel)
══ Ho Chi Minh Trail, supply line
➤ 7th US Fleet ⟶ to FNL
➤ American air raids
I American bombing, dates indicate when most intense

A. 1954–69

▨ FNL control 1965-66
▨ Controlled by Pathet Lao
⟶ Weapons and troops to FNL
● Main US air bases 1969
➤ Tet Offensive, Jan. 30 1968

B. 1969–73

▨ FNL control 1973
▨ Controlled by Pathet Lao
⟶ Amer. blockade May-Dec. -72
▨ FNL base areas in Cambodia

C. 1973–77

➤ FNL and North Vietnamese Final Offensive, March 1975
★ Border incidents 1975-78

Hanoi in ruins after the US bombattack at Christmas 1972

Temple of Borobu-
dur in Indonesia is
quadratic at the base,
515 × 515 feet.

Colossal statue of
Philippine president
(1965–86) Ferdi-
nand Marcos.

A girl from Bali
dancing the traditio-
nal Garuda dance.

183. INDONESIAS INDEPENDENCE

Indonesia 1976. A Dutch colony before independence,
union with Netherl. 1949-56.
Federation of Malaysia,
formed from British de-
pendencies Sept. 16 1963
Earlier boundary

1963 Tense relations with Malaysia. The congress appoints Sukarno
president for life.
1965 A communist coup attempt is
put down. Tens of thousands of
communists, and people of Chinese
extraction, are murdered.
1967 Sukarno is forced to transfer
all of the functions of the president
and the prime minister to general
Ibrahim Suharto. Friendly relations with China are replaced by a
pro-Western policy.
1969 Following a plebescite, Irian
Jaya (map nr. 183) becomes part of
Indonesia, which has administered
the area since 1963.
1975–76 Civil war in the Portuguese area of Timor. Indonesian
forces invade and occupy the territory (map nr. 183).

Above: President Achmed Sukarno
(1901–70) in Manila's bazaars in
the Philippines in 1949. Sukarno, an
engineer, was one of the founders
of the Indonesian Nationalist Party,
PNI, and was imprisoned several
times by the Dutch in the inter-war
period. The Japanese set him free,
and after the Japanese surrender,
Sukarno and Mohammed Hatta
proclaimed Indonesia an independent republic August 17,
1945. Sukarno president. In the
1950s he abolished parliamentary
democracy in the country, instituted
what he called «guided
democracy», and governed the
country in almost dictatorial fashion.
A conflict with the military led to his
political demise in 1967.

Indonesia 1942–76

1942–45 Dutch East Indies is
occupied by the Japanese. An anti-
Dutch nationalist organization, led
by Achmed Sukarno (illus. left),
serves the Japanese in an advisory
role.
1945 Sukarno and Mohammed
Hatta proclaim Indonesia an independent republic on August 17. In
September/October British and
Dutch troops arrive to disarm the
Japanese. Within a short time they
are exchanging fire with Indonesian freedom fighters.
1949 After four years of intense
fighting between Indonesians and
Dutch government forces, both
sides agree to the establishment of
the United States of Indonesia,

within the framework of a Dutch-
Indonesian union.
1950 Local uprisings and continued guerilla activity. On August 17
the Republic of Indonesia is proclaimed, with a strong central
government, led by president
Sukarno,
1956 On August 11 the union with
the Netherlands is finally dissolved. **1957** Military coups, communist infiltration, and rising unrest in
several provinces demanding self-
rule. President Sukarno declares a
state of emergency. Dutch property
is confiscated and Dutch citizens
expelled from the country.

Below: Jawaharlal Nehru's
daughter, Indira Gandhi (1917–
84), put her stamp on Indian
political life for twenty years – both
during the many years she governed
India, and as leader of the
opposition. Her strong personality
made her a natural leader, but made
her many enemies as well.

Mahatma Gandhi as he was depicted after his death in 1948.

The Golden Temple in the Sikhs' holy city or Amritsar.

The Sikhs took up arms following an attack on the Golden Temple in 1984.

Left: Indira Gandhi was shot and killed on October 31, 1984. The pyre was lit by her son, Rajiv, who also succeeded her as prime minister.

India 1939–91

1939–45 British India is automatically involved in the Second World War. Over two million Indians, drafted by the British, take part in the war. Demands for independence are growing steadily.

1947 The huge gap separating Hindus and adherents of Islam is an major obstacle to the creation of an independent, unified India. The country is, therefore, split in two, Hindu India, and Islamic Pakistan (map nr. 184), with the status of dominions. The division leads to the uprooting of many people, and much unrest. Pakistan resists the inclusion of Kashmir, largely Islamic, in the Indian state. Jawaharlal Nehru becomes India's first prime minister.

1948 Mahatma Gandhi, who led the struggle for independence, dies at age 78.

1950 On January 26 India becomes an independent republic with ties to the Commonwealth.

1952 The first election with universal suffrage gives the Congress Party and prime minister Jawaharlal Nehru a resounding victory.

1955 Moscow premises India economic and technical aid.

1959 Border disputes with China, which continue until 1967 (map nr. 184).

1964 Nehru dies and is succeeded by Lal Bahadur Shastri.

1965 August 5, Indian troops enter Pakistan. Cease-fire on September 22, following Soviet meditation, but the situation in the area remains extremely tense.

1966 Indira Gandhi becomes leader of the Congress Party and prime minister.

1971 Civil war in Pakistan. India intervenes actively on the side of East Pakistan, which separates from Pakistan (see col. 4).

1975 Indira Gandhi declares a state of emergency in order to crush a growing, albeit disorganized, opposition. Thousands are executed.

1977 Indira Gandhi loses the election and is succeeded by Morarji Desai of the Janate Party coalition.

1980 The Congress Party makes an impressive comeback. Indira Gandhi becomes prime minister.

1984 Indira Gandhi is shot and killed by two of her own Sikh bodyguards. Gandhi's son, Rajiv, is sworn in as India's new prime minister.

1989 The Congress Party loses the election and Gandhi is succeeded by V. P. Singh.

1991 Rajiv Gandhi is assassinated during the General Elections campaign.

Bangladesh 1947–84

1947 The eastern region of former British India becomes part of the dual state of Pakistan (East Pakistan).

1970 The Awami League, which is agitating for independence for East Pakistan, wins the December elections.

1971 Pakistani government forces enter East Pakistan in February. The Awami leader, Mujibur Rahman, is arrested, and all political activity is banned. In December Indian troops enter the country in support of the liberation forces, and on December 17 the war is over. East Pakistan becomes the state of Bangladesh. 1972 Rahman is set free in January and becomes head of the Government.

1975–77 Rahman is killed in a coup d'état. Zia Rahman seizes power and becomes president (1977).

1981–84 Zia Rahman is murdered. The country is under military rule until General Hussein Muhammed Ershad becomes president in 1984.

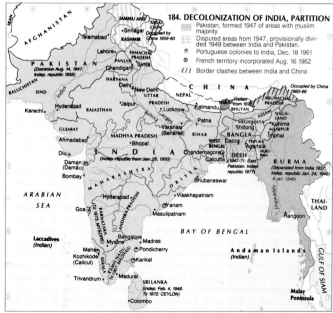

184. DECOLONIZATION OF INDIA, PARTITION

185. EASTERN EUROPE 1945-90

Occupied by USSR 1945
Soviet-bloc countries 1945-1989/90
Iron Curtain
"Stalinist" 1945-61

Revolution in Eastern Europe 1989–90 – communist dictatorships collapse

On Nov 9 1989 the East German authorities opened the Berlin Wall. For the first time since its erection in 1961 East Germans were free to cross into West Berlin. Its demolition climaxed and symbolized the revolutions which in the fall of 1989, had brought down all the communist dictatorships, Moscow's allies. This brought an end to the Iron Curtain, which for over 40 years had divided Europe. Immediately the Soviet Union's former satellites began moving toward western style democracy and a market economy.

In **Poland** the anti-communist revolt had shown omens as early as 1980, when Lech Walesa had founded the free trade union, Solidarity just as a wave of strikes paralyzed the country. Though Gen. Jaruzelski banned the movement and proclaimed martial law in 1981, Solidarity remained a powerful political force. After a new strike wave in 1988 the communist government opened negotiations, which in 1989 led to partially free elections and Solidarity's great victory at the polls and a majority in the new government (Aug. 1989). Though the communist party dissolved itself in Jan. 1990, Jaruzelski, the old regime's leader, remained president until Dec. 1990, when Poland's first democratic presidential election since World War II replaced him with Lech Walesa.

In **Hungary** growing dissatisfaction forced the communist leader, Janos Kadar, in power since 1956, to resign (1988). In Oct. 1989 the communist party dissolved itself and in March 1990 the first democratic elections were held since 1945, and led to the formation of a non-socialist conservative coalition government.

In **Czechoslovakia** peaceful mass demonstrations led by students and intellectuals brought down the communist dictatorship in Nov.-Dec. 1989. The party leadership, headed by Milos Jakes, resigned; and in Dec. Vaclav Havel, playwright and critic of the regime, was elected president. In June 1990 democratic elections gave a great victory to the Citizens Forum, the main force behind the revolution.

Romania had been ruled since 1948 by a communist dictatorship. In 1965 Nikolai Ceausescu had assumed power and made himself the object of an ever more grotesque personality cult, where his Securitate (secret police) exerted a reign of terror. Many people died in the revolution of Dec. 1989, when he was overthrown. A provisional government, the National Salvation Front, took over, and in May 1990 parliamentary elections were won by the Front, whose leader, Ion Iliescu, became president. But the opposition, claiming electoral fraud, accused the Front of being dominated by ex-communists who were sabotaging democratic developments. The atmosphere became even worse after the government (June 1990) had violently unleashed miners against a demonstration, killing five people.

Bulgaria, too, was affected. In Nov. 1989 Todor Zjivkov, the country's communist dictator since 1954, was overthrown. Reforming itself as a 'democratic socialist party', the Communist party won the first free elections (June 1990). But the opposition refused to accept this new government and in the fall of 1990 the country was paralyzed by strikes and demonstrations. In Dec. a coalition government was formed, with the communists (now calling themselves the Socialist party) in a minority. This meant the end of 45 years of communism in Bulgaria.

Lech Walesa

Vaclav Havel

Below: *The hated Berlin Wall, erected in 1961 (see map 167), was opened 9-10 November, 1989, and later torn down.*

Left: *Mikhail Gorbachev. General Secretary of the USSR from 1985, and President from 1988.*

GERMANY from Oct. 3 1990

In East Germany (DDR) a mass flight of East Germans to the west began in the spring of 1989. At the same time the country was paralyzed by pro-democracy demonstrations. As a result of the crisis Erich Honecker, the communist leader, was deposed; and shortly afterward the Berlin Wall was opened. A month later the forced resignation of Honecker's successor, Egon Krentz, put an end to the communist dictatorship. The country's first democratic elections (March 1990) were won by the anti-socialist Christian Democrats on a platform of reunion with West Germany. Economic collapse hastened the progress. In July the D-mark was introduced through monetary union with West Germany, and on Oct. 3 1990 the two German states were politically reunited, the former DDR becoming a member of the NATO alliance and of EEC.

The East European revolutions also reached the other two communist dictatorships on the Balkan peninsula. Ever since World War II YUGOSLAVIA had been ruled by a communist party, but since 1948, under president Josip Tito (d. 1980), it had preserved its right to find its own national path to socialism. From 1989 demands for democratic reforms fused with growing clashes between different ethnic groups. This resulted in a crisis which in 1991 tore the country apart (see p. 132).

After democratic multi-party elections in 1990, power in four federal republics went to non-socialist or nationalist parties. But in Serbia and Montenegro the socialists (i.e. the reformed Communist party) retained power. Slovenia and Croatia declared themselves sovereign states, with their own laws taking precedence over those of the federation. Both threatened to secede. While tension between Slovenes/Croats and Serbs came ever closer to flashpoint,

the country was also shaken by a severe economic crisis, and there were outbreaks of violence in Kosovo, a Serbian province whose Albanian majority demanded greater autonomy. Since 1945 ALBANIA had been governed by a ruthless communist dictatorship. In 1990 the country was shaken by pro-democracy demonstrations demanding the citizen's right to go abroad. Thousands of Albanians began fleeing the country. In Dec. 1990 the communist president Ramiz Alia accepted a multi-party system.

The Communist party, now on a platform of democratic socialism, renamed itself the Worker's Party. In 1991 democratic elections gave it the victory. The democratic opposition showed considerable discontent with election results and violence occurred in many parts of the country. In March 1992 a new election gave victory to the democratic opposition, which formed the first government in 47 years without communists. The communist president Ramiz Alia resigned and the leader of the democratic front, Sali Berisha, became president.

A factor crucial to these East European rebellions was the reforms introduced by the Soviet leader Michail Gorbachev. Coming to power in 1985, he had tried to liberalize the Soviet Union's political and economic system and at the same time had encouraged reformist movements in eastern Europe, and rescinded the doctrine of military intervention against any threat to any communist state's political system. A withdrawal of military support by Big Brother spelled the East European satellite regimes death-knell.

Economic strength and movement towards unity in the West

In contrast to the pattern of economic decay and disintegration in Eastern

Europe the western half of the continent during the 1980s experienced a remarkable economic boom - and also took important steps towards closer cooperation. This development played an important part in the great transformation of Europe that began in 1989. The organizations for East European collaboration collapsed. The Warsaw Pact, created in 1955, was abrogated in 1991, and Comecon, Eastern Europe's economic organization (see map on page 168), also disintegrated. Western Europe's political and economic institutions - above all the EEC - became the model for economic reforms in the East. Most of the new democratic governments in Eastern Europe wanted to be integrated in the western economic collaboration as soon as possible. But the road into EEC remained a long one. The western EEC-states at first wanted to strengthen their internal collaboration. As a first step they decided to create a free internal market. This "Europe without borders" began to be realized in January 1993. In 1991 the EEC-states signed a treaty in Maastricht with the aim of developing a tighter economic and political union by the end of the century. This scheme ran into severe difficulties however. In 1992 Denmark said no to the Maastricht treaty in a referendum. Opposition to the treaty also grew elsewhere in the Community. A serious currency crises the same year threatened the plans of shaping a single European currency by 1999. On top of that the western countries suffered from a deep economic recession, which in 1990 replaced the big boom of the 80s. The prospects of uniting Europe into a kind "United States of Europe" looked gloomy (at least for the time being) in the beginning of 1993. Those difficulties in the west at the same time dampened the chances for Eastern

EUROPE at beginning of 1991

Member of EEC

/// Member of EFTA

Member of NATO

Belongs to Council of Europe 1990-91

Europe to be incorporated in the western sphere of economic integration any time soon.

The USSR, a Disintegrating Empire

1. Estonia
2. Latvia
3. Lithuania
4. Belarus
5. Ukraine
6. Moldova
7. Georgia
8. Amenia
9. Azerbaijan
10. Turkmenistan
11. Uzbekistan
12. Tajikistan
13. Kirgizia

● Areas where ethnic and/or political conflicts have caused loss of life since 1986.

By early 1990 all 15 Soviet republics had declared their sovereignty with their own laws taking precedence over those passed by the Soviet parliament. Estonia, Latvia, Lithuania and Georgia went furthest in striving for complete autonomy. They had expressed a desire to leave the Union altogether. For development from 1991 see p. 132.

THE DISCORDANT YUGOSLAVIA

AUSTRIA
HUNGARY
SLOVENIA
CROATIA
ROMANIA
BOSNIA & HERCOVINA
SERBIA
ALBANIA — MACEDONIA
BUL-GARIA
GREECE

Ethnic Moslems

Serbs

/// Croats

Slovenes

Monte-negrins

Albanians

Macedonians

Hungarians

International boundary

Republic boundary

Regional boundary

USA - World Power

After World War II the USA was seen to be one of the world's two superpowers. It abandoned the isolationism which had long determined its policies toward the world outside the American continents. More than, anything else it had been USA's economic and military strength which had secured the Allies' victory over Germany and Japan. The American nation became conscious of its own power and of having a global mission: to defend western-style democracy against communism.

Through the NATO alliance, led by the USA, a barrier was raised against the military threat from the USSR, the other superpower. By a chain of alliances and military bases the USA also tried to create guarantees against communist expansion elsewhere. By intervening in the Korean War (1950-53) it repulsed communist aggression

(see p 119). The USA's next major military effort was in Vietnam (Indochina), where it fought a sanguinary war 1964-73, the first not to end in an American victory. After the US army's withdrawal, South Vietnam was conquered by North Vietnam, and communist regimes also took power in the rest of Indochina. This defeat shook American self-confidence. Though the USA retained its leading role among democracies, its total dominance seemed to have shrunk. A factor here was that the USA's relative weight in world economy dwindled as new economic forces made themselves felt - primarily Japan and the EEC. This development has continued in the early 1990s. At the same time revolutionary events have reinforced the USA's relative global weight and prestige.

The fall of communism in Eastern Europe and above all the death of the

USSR have left the USA as the sole remaining military superpower - a position that became obvious in connection with the liberation of Kuwait and the total American-Allied

victory over Iraq in 1991. The alliance, sanctioned by the UN, was dominated and led by the USA, which also contributed to 80 % of the war effort.

1991

USA IN THE WORLD

● Air base   ▲ US military force (over 10,000 all arms) outside contiguous USA

× Naval base

LATIN AMERICA

1. GUATEMALA
2. BELIZE 1981 x
3. HONDURAS
4. EL SALVADOR
5. NICARAGUA
6. COSTA RICA
7. PANAMA
8. JAMAICA 1962 x
9. BAHAMAS 1973 x
10. ST. KITTS-NEVIS 1983 x
11. ANTIGUA-BARBUDA 1981 x
12. DOMINICA 1978 x
13. ST. VINCENT 1979 x
14. BARBADOS 1966 x
15. ST. LUCIA 1979 x
16. GRANADA 1974 x
17. TRINIDAD-TOBAGO 1962 x

(1981 = year of liberation,
x = member of CARICOM)

Members of:

LAIA, Latin American (economic) Integration Association, 1981, successor of Latin American Free-Trade Area founded 1960.

CARICOM, Caribbean Community for co-operation in economic, foreign policy and other areas, founded 1973.

CACM, Central American Common Market, founded 1960.

Production of Narcotics

△ Cocaine   □ Marijuana
○ Opium
● Important Centre of Narcotics Refining

US Intervention in Central America:

Guatemala, military intervention leading to establishment of conservative regime in 1954;

EL SALVADOR, many years of military and economic aid to government fighting left-wing insurgency:

HONDURAS, financing of bases for 'contras' gerillas in 1980's:

NICARAGUA, military and economic aid to 'contras' fighting the left-wing regime, until 1990;

PANAMA military intervention to overthrow the dictator Noriega in 1989;

GRENADA, military intervention following power struggle within the Marxist regime in 1983 US-friendly interim government installed

Latin America was largely decolonialized in the early 19th century (see pp. 94-95). The 1960-1990 period brought a new anti-colonial wave: 12 Brit. colonies and the formerly Dutch Surinam became independent. The map dates the independence of each country since 1960. The Latin-American countries have set

up several organizations for economic (and to some extent also political) collaboration. But trade within the region is still poorly developed. Only 14 % of Latin-American trade is internal to the region (1989) - the rest mainly with USA and Europe.

Democracy on the March

In the late 1970s most Latin-American countries had been overt or indirect military dictatorships. A major change occurred from 1979-90. Most of the dictatorships were replaced by elected civilian governments. This re-democratization process began in Ecuador, where the military regime was dismantled, with democratic elections in 1979. Since then democracy has been established in Peru (1980), Bolivia (1982), Argentina (1983), Brazil, Guatemala and Uruguay (1985), Panama (1989), Chile and Nicaragua (1990). After the dictator Stroessner's fall in 1989 Paraguay, too, has begun to become more democratic. For many decades Central America has been ravaged by civil wars and sanguinary political violence, conflicts aggravated by great power interference. The USA's traditional policy of naked power politics, together with its immense economic influence, has determined much of the region's development. Up to the late 1980s the USSR, helped by Cuba, was aiding and abetting left-wing revolutionary movements. In 1987 the Central-American countries agreed on a peace plan that made possible the 1990 transition to democracy and an end to civil war in Nicaragua. In 1992 a UN-mediated peace agreement between the government and the FMNL guerilla in El Salvador made it possible to put an end to 13 years of civil war and to prepare for democratic elections.

Democracy under threat

The military still exerts great influence in Latin America. Many of its democracies must take military interests into account to avoid new military coups. This has thwarted attempts to bring to trial earlier military dictatorships' criminal infringements of human rights. After the wave of democratization there have been setbacks in Latin America in the early 1990s. In Haiti the first president chosen in free elections was installed in 1991. But this experiment in democracy lasted only a few months before the president, Jean-Bertrand Aristide, was deposed in a new military coup. In Peru the democratically elected president Alberto Fujimori staged a military coup of his own in 1992, suspending democracy with the motive to fight corruption and to give the country a strong leadership in the war against leftist guerillas. In Venezuela the military in 1992 staged two military coup attempts against the democratic government. One of the big problems has been the burden of debt incurred by most Latin American governments in the 1970s. Latin America as a whole today has the world's heaviest debt per capita. Many governments are tied to restrictive economic policies demanded by the World Bank as a condition for further loans. In combination with massive inflation and flight of capital this represents a constant risk of social unrest and a latent threat to democracy. Another threat is the drug traffic. Cocaine, opium and marijuana are produced in large areas of Latin America (see map).

The Middle East

The Middle East is one of the world's most conflict-ridden regions. In 1988 the eight-year war between Iran and Iraq ended with a suspension of hostilities. Neither side had gained anything. The war had cost both countries a million lives and immense material devastation. Yet Iraq was still militarily powerful. And its president Saddam Hussein did not relinquish his aggressive policies. Kuwait became his next target. He hoped by control of this small but oil-rich neighbor to solve his economic problems and realize his dream of becoming the leader of The Arab World.

On Aug 2 1990 Iraq occupied Kuwait. But world reactions were much stronger than Saddam Hussein had anticipated. The great powers of the

UN agreed to apply economic sanctions. Forming a military alliance the USA sent large military forces to the Persian Gulf area. The UN Security Council empowered the alliance to attack Iraq unless it had evacuated Kuwait by Jan. 15 1991. On Jan. 17 the allies launched devastating air attacks on Iraq's industrial, military and communications system, lasting for six weeks. A final land offensive, based on total command in the air and an outflanking movement, liberated Kuwait. On Feb 28 hostilities were suspended. The brief war had cost Iraq more than half her army. Despite this total defeat, Saddam Hussein remained in power, with enough forces to suppress a revolt by Shiite Muslims in the south and Kurds in the north. This civil war unleashed a further catastrophe.

Half of Iraq's 3.5 mill. Kurds fled toward Iran and Turkey. Later, with temporary protection from the UN-alliance, the Kurds have been able to establish de facto self government in parts of Iraqi Kurdistan.

Kuwait had been liberated; but Israel remains in occupation of the West Bank and Gaza strip, where the Arab population has been in revolt since 1987, demanding an independent. Palestinian state. After the Kuwait war, the USA has launched a diplomatic offensive to seek a peaceful solution to this conflict. If it fails, new wars threaten the Middle East.

South Africa

Great changes occured in southern Africa from 1989-91. In 1989 South Africa relinquished its control of Namibia, which in 1990 became an independent state, with a democratically elected black government. At the same time South Africa moved swiftly towards equal rights for its black majority. Politically, socially and economically the black population had been discriminated against by race laws, first introduced in 1948. As the oppression became worse in the 1970s and 80s, many countries decided to exert pressure on South Africa in the form of economic sanctions. When widespread unrest broke out among the black popula-

THE MIDDLE EAST
The Political Scene in the Middle East after the Gulf War

Arab countries that joined US-led alliance against Iraq. Post-war security cooperation makes them the leading political bloc in the region
Gulf Cooperation Council, an economic and political organization dominated by Saudi Arabia
Turkey. US-ally through membership of NATO
Defeated Iraq:
Occupied by allies in February 1991
Revolts in March 1991, controlled at times by Shia Muslims (in south) and Kurds (in north)
Countries sympathetic to Iraq during war
Shiite and Kurdish refugees (March-April 1991)

SOUTHERN AFRICA

Of South Africa's 32 million inh. 68% are black, 18% white, and the rest Indians or "colored" halfbreeds.

South Africa was formed in 1910 by a fusion of the British colonies of Natal, Cape Procince, the Orange Free State and Transvaal. Remained in the British Commonwealth until 1961.

tion in 1985, the governing Nationalist party realized the system must be reformed. But not until F. W. de Klerk became president in 1989 did such reforms really begin. In 1990 the government released the ANC leader Nelson Mandela from his long imprisonment. ANC became legal, and peace negotiations were opened with Mandela for the abolition of apartheid.

The government abolished several basic apartheid laws in 1991. That same year negotiations started for a new constitution with democratic elections and a black franchise (CODESA - Conference for Democracy in South Africa). The peace process has been slow with many disturbances and violence between ANC and rival black groups. There have been accusations from ANC that the white government encourages the violence and is attempting to bargain for more white political influence than is justified by their numbers. There is also resistance and violence from rightwing white groups who are opposed to the idea of giving any political rights to the black majority.

Oceania

Oceania consists of 25,000 islands, scattered over a region equal to one-tenth of the earth's surface. Here the domination of the western colonial powers persisted longer than in any other part of the world. Australia and New Zealand apart, all Oceania's independent states have come into existence in the last 25 years.

For a long while Oceania remained politically stable. Only in recent years have its islands experienced troubles, often related to ethnic conflicts. A revolt broke out among the Kanakas

in French Caledonia in the late 1980s; Fiji was hit by two military coups in 1987; and the island of Bougainville, in Papua-New Guinea, has been in a state of revolt and demanding its independence since 1989. The South Pacific Forum (founded 1971) is the region's organ for political and economic collaboration among all its 11 countries. In 1985 SPF declared the whole region a non-nuclear zone, thus defying France, which since 1966 had been carrying out nuclear bomb tests on Mururoa.

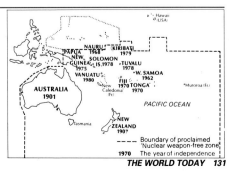

Boundary of proclaimed 'Nuclear weapon-free zone'
1970 The year of independence

ICELAND
SWEDEN
NORWAY
FINLAND
RUSSIA
ESTO-
NIA
LATVIA
IRELAND
DENMARK
LITHU-
ANIA
UNITED
KINGDOM
NETHER-
LANDS GERMANY
BELARUS
POLAND
BEL-
GIUM LUX.B.
CZECH
UKRAINE
REP.
SLOVAKIA
MOL-
FRANCE
AUSTRIA
DOVA
SWITZ.
HUNGARY
ROMANIA
1 2
PORT-
BOSN.
UGAL   SPAIN   HERZ. 3
ITALY
ALBA- 4
NIA
GREECE
1. SLOVENIA
2. CROATIA
3. YUGOSL.
4. MACEDONIA

EUROPE 1993 (April)

Eastern Europe 1991–93 - ethnic strife, bloody break-up of nations and economic chaos dominate the scene.

The dissolution of the Soviet Union in 1991 has been the most revolutionary event on the world scene in the past 40 years. The Soviet empire still was one of the two superpowers at the start of the 1990s. Its rapid break-up has profoundly changed the old world order to which humanity had become accustomed since the end of the second world war. The first cracks in the empire appeared when it be came evident that president Mikhail Gorbachev was unsuccessful with his reform policy that was launched to save a crumbling economy. At the same time his political scheme for more political openness, *glasnost*, inspired nationalist forces in many Soviet republics to ask for more authority to handle their own affairs. This development reinforced the conservative opposition against the president.

An attempt to depose president Gorbachev was made in August 1991. Behind it was a group of conservative communists who had wished to restore the old order and keep the Union together. The failure of the coup produced the opposite result. Boris Yeltsin, president of Russia, emerged as the leader of the democratic forces and as the predominant leader in the Union and made the restored Soviet president Gorbachev powerless. The

communist party was declared illegal. The authority of All-Union institutions collapsed. Before the end of the year all 15 former Soviet republics had declared themselves independent. The USSR no longer existed and president Gorbachev resigned on Christmas Day of 1991. Most of the former Soviet republics (except the Baltic states and Georgia) agreed to a limited degree of cooperation within a loosely knit **Commonwealth of Independent States (CIS)**. CIS was later abandoned by Moldova and Azerbaijan, leaving nine states within the organization. Most of the states emerging from the ruins of the USSR have, in principle, favored the idea of building democracy and a free market economy. In reality this process has been hampered by deep economic crisis and in many places by the old communist élite striving to keep their traditional power. Many of the newborn states are plagued by wars, due to ethnic antagonism and disagreement about borders. A civil war started in 1992 in Tajikistan. Repeated fighting has occurred in Moldova between Moldovians and the Russian minority, which demands self government. A five year old war continued in 1993 between Armenia and Azerbaijan about the Azeri region of Nagorno-Karabach, which is claimed by the Armenians. Georgia has been ravaged by civil wars since the start of the 1990s.

The struggle for democracy and economic revival in the ex- Soviet territories very much depends on what happens in Russia, which was

the predominant part of the Soviet Union. When Russia became a state of its own, president Boris Yeltsin vowed to lead the country towards market reforms and democracy. One year later his reforms, however, had made no progress - the only result so far being rapid inflation, chaos in the administration and worsening living conditions for the Russian people. Yeltsin's authority was challenged by a strong opposition in the mainly conservative parliament, which tried to slow down the reform process, – and even put an end to it. The power struggle between president and parliament paralyzed Russia in the beginning of 1993. At the same time many Russian "autonomous republics" and other regions on the periphery of the country showed signs of independence, turning their back on the central government in Moscow.

Yugoslavia, like the Soviet Union, has been torn apart as a result of the anti-communist revolutions in Eastern Europe (for events before 1991, see page 129). In the summer of 1991 two republics in the former federation of Yugoslavia declared themselves independent – Slovenia and Croatia. This was the beginning of a series of wars in the former Yugoslavia. The federal government in Belgrade tried to uphold its authority by ordering the federal Yugoslav army to intervene in Slovenia. This action however was brief, ending with retreat of the federal troops after a few weeks. The fighting became worse in Croatia, which has a considerable minority of Serbs, most of whom were opposed to the prospect of having to live in a state dominated by Croats. This war was as much a fight with the central, Serb-dominated government in Belgrade as a civil war between Croats and Serbs inside Croatia. After bloody fighting a fragile peace was created with a UN-mediated cease-fire in January 1992. Slovenia and Croatia gained international recognition as independent states. But vast areas of Croatia with a Serbian majority remained outside the Croatian government's control. In April 1992 another former Yugoslav republic, –Bosnia-Herzegovina, – was inter-nationally recognized as independent. This started a new war between the dominant ethnic groups – Serbians, Croatians and Bosnian Muslims. The Serbs, and later the Croats, proclaimed their own independent areas in Bosnia. In the following year the Serbs conquered 70% of the country in bloody fighting, inflicting terrible suffering on the civil population. The Muslims were

systematically killed, put in prison camps or driven away from vast areas in a systematic campaign of "ethnic cleansing". War crimes were committed by all adversaries in the war. The world, however, condemned the Serbs as the most responsible party, in particular the regime in Serbia, lead by the fanatic Serbian nationalist (and former communist) Slobodan Milosevic. Serbia's involvement in the war led to a UN decision to isolate "mini-Yugoslavia" (Serbia-Montenegro) with an economic blockade. At the same time the UN tried, in cooperation with EEC, to enforce a peace plan. Still, in March 1993 peace diplomacy had produced no tangible results. Pressures on the Western powers grew to stop the war by military intervention.

Another former communist country, Czechoslovakia, dissolved in the wake of the rising tide of nationalism in Eastern Europe . After elections in 1992 a leftist and nationalistic party came into power in Slovakia, the eastern part of the federation. The Slovak prime minister Vladimir Meciar was unable to agree with his colleague in the Czech federal republic, Vaclav Klaus, about the terms for preserving the federation. Instead they agreed to divide the country into two separate states. On January 1st 1993 the Czech Republic and Slovakia were formed. Fortunately this partition was peaceful. The former president of Czechoslovakia, Vaclav Havel,was nominated as the first president of the Czech Republic.

ABBREVIATIONS USED IN MAPS AND TEXT

Aug.	*August*	Fr.	*French, France*	PR	*Principality*
Beg.	*Beginning*	GR DUCHY	*Grand Duchy*	Nov.	*November*
Brit.	*British*	indep.	*independent*	Port.	*Portugal,*
Ca.	*Circa*	internat.	*international*		*Portuguese*
Chin.	*Chinese*	Ital.	*Italian*	Pruss.	*Prussian*
Denm.	*Denmark*	Jan.	*January*	Prot.	*Protectorate*
Dec.	*December*	Jap.	*Japan*	REP	*Republic*
DSP	*Despóty*		*Japanese*	Russ.	*Russia, Russian*
Egypt.	*Egyptian*	KGD.	*Kingdom*	Sept.	*September*
ELECT.	*Electorate*	LGR	*Landgravate*	Sp.	*Spain, Spanish*
EMP.	*Empire*	MGR	*Margravate*	Sw.	*Swedish*
Febr.	*February*	Oct.	*October*	Turk.	*Turkey, Turkish*

MAP INDEX

INDEX (SUBJECT AND NAME)

HISTORICAL TIMELINE

Since the earliest beginnings in Africa, the evolution and
spread of humankind has seen great variations.
For different reasons at different times highly developed
cultures have evolved in one area of the world, while people
in other areas were still living as hunters and gatherers. By the
same token, it has often happened that geographically distant
cultures have seen similar achievements in such areas as
tool-making, art, architecture and music at approximately the
same point in time.

With the aid of modern scientific methods archaeological
finds can now be dated with great accuracy. However, there
is often much disagreement over the conclusions to be drawn
from these finds. And besides, archaeologists have yet to
explore and research many areas of the globe. And that is why
our knowledge of the earliest ages of human evolution are
constantly changing and being reevaluated. The overview on
the following pages shows, in chronological order, examples
of historical events and human development at approximately
the same moments in history around the world.

	UNITED STATES	AFRICA	ASIA	AUSTRALIA
		The earliest evidence of tool-making in Eastern Africa. Chopping and crushing tools.		
		Palaeolithic: Old Stone Age (Hunter-gatherer Stone Age).		
		Homo sapiens sapiens becomes the dominant species.		
180 million BC	During the Mesozoic Era, dinosaurs roam on the North American Continent.	Large areas of Africa inhabited.	Large areas of Asia inhabited.	Large areas of Australia inhabited.
10,000 BC	Groups of Paleo-Indians cross the Bering Strait from what is now Siberia to Alaska.	The total human population of the earth increases to ca. 2 million.		
8,000 BC	The Atlatl, or spear, is widely used in the Southwest. It revolutionizes hunting techniques.	Hunter-gatherer cultures evolve, having roughly the same characteristics in all parts of the world. During points, knives and scrapers), masters the controlled use of fire, and makes practical use of stored energy		
		The Neolithic: New Stone Age (agricultural Stone Age). Also known as the Neolithic Revolution. The of food. In agricultural areas, the population grows rapidly.		
5,000 BC	Corn is domesticated in Mexico and spreads from there into the Southwest.	Man begins to cultivate and refine certain wild plants – the birth of our food grains. The taming and This gradual change in the method of food-production leads to a change in the nomadic life-style of		
2500 BC	Agriculture and pottery-making begin in Eastern North America.	and begin building permanent settlements. A consequence of grain production is the need for storage archaeologist.		
		The more-or-less unilinear evolution of world culture is broken, as development quickly accelerates in		
100 BC	Farming villages appear in the Southwest.	Evidence of food grain production in the Nile Valley	In the eastern Mediterranean and Mesopotamia people begin using copper and building houses of clay brick.	
			Irrigation of fields begins in Mesopotamian and Indo-Iranian areas. Advanced pottery-making. Fertility cults.	
		The beginnings of so-called high culture in various parts of the world. Increased agricultural production elite evolves, and develops the art of writing for administrative purposes. The wheel and the seal are applications. Religious beliefs find expression in monumental constructions.		
		Egypt – glazed ceramics. Bronze production. The first flutes and harps.	In the Tigris-Euphrates area the Sumerians invent the wheel and the cart. They also make use of the potter's wheel. China – cultivation of rice begins in Yang Xao. Thailand – copper tools.	Aborigines make use of musical instruments – «didgeridoos and *bullroarers».
		The first Egyptian dynasty ca. 3000–2810 BC. First Egyptian cities of brick.	The Sumerians develop writing – cuneiform symbols on clay tablets. Early Indus Valley culture from ca. 3000 BC. Cultivation of cotton and use of coloured textiles.	
		Ca. 2630 – The first pyramid at Saqqara-worlds oldest stone structure.	Large cities with high standards of hygiene. Bathrooms in villas and apartments. Utilization of metals.	
		Calendar with 365 days. Bantu tribes in West Africa begin migrating to the east and south.		

EUROPE	NORTH AND CENTRAL AMERICA	SOUTH AMERICA		
				2 500 000 BC
				35 000–8 000 BC
				35 000 BC
Large areas of Europe inhabited				
				25 000 BC
this hunter-gatherer Stone Age, mankind learns to make specialized tools (needles, arrow and harpoon (bow and arrow).				8 000–5 000 BC
most important technical advancement in human history takes place during this period – the cultivation		

Dwelling ca. 20 000 BC. | |
raising of certain animal species is also begun (wild sheep, goats, oxen and boar).				
the hunter-gatherer cultures. In order to cultivate plants and animals, human societies stop wandering vessels. Pottery-making flourishes, and is an inestimable source of information for the modern				
agricultural areas.				
Artistic development: Cave paintings in Altamira (Spain) and Lascaux (France) Stone and bone sculpture (Venus of Willendorf) in Eastern and Central Europe. Cultivation of grain along the coast of the Mediterranean. The oldest Scandinavian rock carvings depicting animals. Stone chamber graves, monoliths and temples of stone in Scandinavia, Northern Germany, Holland, Belgium, France (Carnac), Portugal, Spain and Great Britain (Stonehenge) bear witness to a developing religious imagination.	North and Central America inhabited.

Evidence of maize growing in Mexico. | |

Venus of Willendorf.

Dwelling made of mammoth tusks. | |
| due to irrigation makes it possible to feed the populations of growing cities. A political and religious invented. Commerce increases. The alloy we call bronze is used in artistic as well as practical | | | | 5 000 BC
4 000 BC |
| Minoan culture begins on the island of Crete. Advanced tools and weapons of bronze. Cretan ships in the Mediterranean. | | |

Australians making music. | |
| The loom is invented. Early rock carvings in southern Norway. On Crete stone is used as building material. | Central America 3 372 BC – the first year in the Mayan calendar – devised by the priests. | Peru – the first cities and villages along the coast.

Cultivation of cotton begins. | | 3 100 BC |
| | | |

Sumerian chariot. | 2 630 BC |

	UNITED STATES	AFRICA	ASIA	AUSTRALIA
BC – 500 AD	Ceremonial earthworks and burial mounds are built in the Northeast.	Egypt – The Old Kingdom – 2665–2155. The hieroglyphs are already developed. All power, both human and divine, is concentrated in the hands of Pharaoh – at the head of a religious and administrative hierarchy. The characteristic relief and painting is highly developed.	The first Aryan invasion of the Indus Valley region. Bronze used in China.	
400 AD	Pottery is manufactured in the Southwest; Basket-making cultures also flourish in the Southwest.			
950	Anasazi people construct the Mesa Verde cliff dwellings in the Southwest.		1848 – Sumerian decline. The first Babylonian dynasty. The laws of Hammurabi reflect a carefully regulated economy. Advanced system of counting, with 60 as its base – our hours, minutes and seconds. The day is divided into two 12-hour periods. The zodiac devised.	
1000 – 1500	Temple mounds and towns are built in the Mississippi Basin.			
1000	Leif Ericsson explores the Atlantic Coast of North America.		1 800 – Aryans destroy the Indus cultures. Smelting of iron. The caste system develops.	
			1 400 – Use of iron reaches western Asia.	
		1 290–1 224 BC Egypt, under Ramses II expands into Asia. Temple built at Abu Simbel.	Ca. 1 230 – Moses leads the Jews out of Egypt – the Exodus. On Mount Sinai he receives the Ten Commandments from God.	
		Horse and wagon used on trade routes in the Sahara.	1 100 – The first Chinese lexicon.	
		Ca. 600 BC – Bantu people spread in East Africa.	Ca. 1 015–930 BC Palestine – Saul introduces monarchy. David and Solomon reign. Jerusalem becomes the capital of a united Israel. A temple is being built where the Arc of the Covenant with the laws is kept.	
1492	Christopher Columbus sails to the New World.			
1513	Ponce de Leon explores what is now Florida.		Japan 600 BC According to tradition, the empire is founded.	
1539	Hernando de Soto travels in what is now Florida, Georgia and Alabama, reaching the Mississippi River.		586 – The Babylonians occupy Jerusalem, destroy the temple. The Arc disappears. Part of the population taken away – »the Babylonian Captivity.«	
1540	Vasquez de Coronado explores what is now Arizona and New Mexico.		The Phoenicians – a civilized trading people – found a colonial empire in the Mediterranean (f.ex. Carthage). Develop written language.	
1550	Great Plains Indians ride horses to hunt buffalo.			
1607	Captain John Smith, an Englishman, helps establish Jamestown, VA.		565 – Buddha (Siddharta Gautama) is born and creates the first missionizing world religion of nonviolence. Buddhism's temple architecture develops.	
1610	Henry Hudson reaches what is later called the Hudson Bay and River; Spanish settle Santa Fe.			
1619	The first Black slaves are brought to Virginia.		China – 6th century Confucius preaches reverence for ancestors, loyalty to authority and respect for ancient customs.	
1620	The Pilgrims arrive in Cape Cod, settle Plymouth Colony and draft the Mayflower Compact.			

EUROPE	NORTH AND CENTRAL AMERICA	SOUTH AMERICA		
The Bronze Age commences in Europe. Crete – the first palace of Knossos.			Mural from Knossos.	2000 BC 1830 BC
2000–1400 – Crete dominates the Aegean. Minoan culture flourishes – cult of the bull and bare-breasted goddesses.				
Ca. 1400 – Mycenae – local upswing. The «golden» Mycenae. Fortress with the Lion Gate. 1200–1100 The Trojan War. Earlier advanced cultures have meant little to recent developments. In the last millennium BC cultures arise in Greece and Italy which have a decisive influence on the development of the West. Greece – 776 The first Olympic games. Italy – 753 according to legend, Romulus and Remus founds the city of Rome. Ca. 700–500 The archaic age of Greece. Homer author of the Iliad and the Odyssey. Rise of the Greek city-states. Italy – 700–500 Etruscan civilization flourishes in present-day Toscany. Remarkable artworks – sculptures and murals in grave chambers. Greece – 500–338 The classical period. Art, science and philosophy flourish. 500–449 The Persian Wars: Greek victories at Marathon (490) and Salamis (480). 447–438 The Parthenon is built. Socrates 469–399 – sought by means of leading questions to guide men towards knowledge and proper behaviour. Sentenced to drink cup of hemlock at age 70. Plato 427–347 – The visible things are only reflections of the invisible, eternal ideas. Those who could see the ideas – the philosophers – should govern. Aristotle 384–322 Antiquity's most versatile thinker. The authority in nearly every area of the sciences.	Ca 800–400 Central America. The Olmec culture evolves west of Yucatan. Writing, and a system of counting which includes zero.		Hammurabi standing in front of the sun god. Dead man's heart is weighed. Olympic sprinter. Greek amphitheatre. The Acropolis in Athens.	1500 BC 1200 BC 1100 BC –400 BC

	UNITED STATES	AFRICA	ASIA	AUSTRALIA
			Ca. 300 – Construction begins on the Great Wall of China. China – **206 BC** Han dynasty comes to power initiates a period of intense expansion. Class of Mandarins develops. Confucianism spreads.	
1629	The Massachusetts Bay Colony is founded.			
1639	The first New World newspaper is printed in Cambridge, MA.			
1649	Maryland Assembly grants religious freedom.		**4??** – Jesus born. **30??** – The Crucifixion	
1682	Sieur de LaSalle claims the Mississippi River area for the French.		**Ca. 50** – The early Christians decide that the teachings of Jesus are not meant for Israel alone. Christianity belongs to the world.	
1692	At least 19 people (most of them women) are executed as witchcraft hysteria spreads through Salem, MA.		**60–70** – The great missionary, Paul, dies after founding Christian communities in Asia Minor and Greece.	
1712	A revolt by slaves in New York leads to the execution of 21 Blacks; 6 slaves commit suicide.	Egypt **Ca. 150** Alexandrian philosopher Ptolemy publishes his great geographical work; the earth is the centre of the universe. The geocentric world-view.	**74–94** The Chinese open the Silk Route to the West.	
1732	Benjamin Franklin publishes Poor Richard's Almanac.		**325** – The Council of Nicea determines that even women have souls. Japan **400s**: Immigration from China is major cultural influence.	
1733	Georgia, the last of the original 13 colonies, is founded.	**200** – Camels introduced from Asia and become important means of transport.		
1741	13 Blacks are hanged, 13 burned and 71 deported after a second New York slave uprising.	**439** – The Vandals – a germanic tribe – invade Northern Africa via Spain (Andalusia). **455** – Vandals sack Rome. It is from them we have the term 'vandalism'.	**570–632** Mohammed, founder of the monolithic world religion Islam. M:s revelations, the Koran is Muslims' holy book.	
1770	British troops kill five townspeople in what became known as the Boston Massacre.		China **618–907** Tang dynasty. Political and cultural Golden Age.	
1773	East India Company's cargo of tea is thrown overboard at the 'Boston Tea Party', a colonial revolt against taxes.		**622** Mohammed expelled from Mecca – point of reckoning for the Islamic era.	
1774	The first Continental Congress is held in Philadelphia, PA.		Holy war begun to spread teachings of Islam.	
1775	The American Revolution begins with battles in Lexington and Concord, MA.		**8th century**: Chinese landscape painting evolves. Golden Age of lyrical poetry; court poet Li Tai-po	
1776	The colonists adopt the Declaration of Independence.			
1777	General George Washington defeats Lord Cornwallis at Princeton, NJ; the 13 colonies unite under the Stars and Stripes flag said to have been designed by Betsy Ross.	**700** Bantu tribes cross the Limpopo into Southern Africa and introduce the use of iron. Arabs conquer Tunisia.	**711** Muslim empire from Spain in the west to China in the east. **750** The Arabs learn the art of paper-making from the Chinese.	
1778	British pull out of Philadelphia, PA; Congress prohibits importing slaves.	The Berbers are driven out of Ghana, and the country comes under black African rule.	Japan **880s**: Japan becomes a feudal state with cultural blossoming (the novel, and philosophical texts, textiles and metallurgy, masterpieces in lacquer).	
1783	The British and the United States sign a peace treaty.	**800** – Trade develops across the Sahara between North and West Africa. Camels of major importance.		
1787	The US Constitution is adopted.	**850** – The Great Citadel in Zimbabwe is built.		

EUROPE	NORTH AND CENTRAL AMERICA	SOUTH AMERICA		
338 – The Macedonians defeat the Greeks. The Hellenist Age begins. Alexander the Great spreads Greek culture eastward through his conquests. Founds Alexandria in Egypt, which becomes a centre of culture and commerce. Italy – The Etruscans are expelled. Rome becomes republic. Archimedes **287–212** – epoch-making contributions to arithmetic, geometry, mechanics. **60** – Julius Caesar in power. **44 BC** – Caesar murdered on the 15 March. Augustus wins power-struggle after Caesar's death. Absolute monarchy from 27 BC. Pax Romana 200 years of relatively peaceful development and consolidation. **64 AD** – Rome burns. Emperor Nero blames the Christians. Persecutions. Italy **79** – Vesuvius erupts, destroying the cities of Herculaneum and Pompey. **122** – The Roman Empire spans the Mediterranean, Western Europe and England. Emperor Hadrian has a fortified wall built against the Scots. **337** – Constantine, the first Christian emperor. Attila, The Hun invades Europe. **493** – Theoderic the Great, king of the Ostrogoths, conquers Italy. The West Roman Empire ceases to function. France Clovis (d. 511) founds the Frankish kingdom (approx. present-day France) and becomes the progenitor of the long-haired Merovingians' royal line. Paris becomes the capital. French evolves from Latin. **500s** – Gregorian church music develops. Spain 711 Arabs (the Moors) begin their invasion of Spain. The Arabs establish the caliphate of Cordoba. France **768** – Charlemagne, Frankish king, emperor in 800, expands the power of both church and crown. Church building on a vast scale, bishoprics created. Carolingian renaissance; classical works, bible manuscripts, etc., are collected, copied and illustrated. The first real illustrations.	**100–600** – Mysterious culture around Teotihuacán in Mexico with monumental architecture. **300–900** – Mayan culture on the Yucatan peninsula. Pyramids and murals. Advanced knowledge of astronomy; calendar. Cultivation with crop-rotation.	Peru **200 BC** – The Mochica civilization develops; large irrigation system in the desert.	 Roman warrior from 300 BC. Roman coach. Camel on the Silk Road. Mohammed preaching. Roman ship taking on grain. Charlemagne's palace.	**300 BC–0** **0–800 AD** **800 AD**

	UNITED STATES	**AFRICA**	**ASIA**	**AUSTRALIA**
1789	George Washington becomes the first US president; John Adams, vice president; Thomas Jefferson, secretary of state; and Alexander Hamilton, secretary of the treasury.			
1790	Congress meets in the temporary capital of Philadelphia, PA; the US Supreme Court holds its first session.			
1791	The Bill of Rights goes into effect.			
1792	Alexander Hamilton and John Adams form the Federalist Party.			
1793	Eli Whitney files for the patent on the cotton gin.	*961–1171* – The Fatimid caliphate in Egypt is at the centre of a Shiite attempt to take over Islam. The Fatimids gained control over the lucrative trade between southern and eastern Asia. Their dominance continues on until the Europeans begin sailing around Africa.	*960–1279* – China. The Sung dynasty comes to power and contact with the West is broken. Period of blossoming in the ceramic arts. Gun powder is used for military purposes.	
1797	John Adams becomes the second president.			
1800	Washington, D.C., is chosen as the United States capital.			
1803	The US buys the Louisiana Territory from France for approximately $15 million;			
1804	Lewis and Clark begin their expedition of the Northwest.			
1807	Robert Fulton makes the first steamboat trip from New York City to Albany, NY.			
1808	Slave importation is outlawed.			
1812	After the British seize US ships and give Native Americans guns to intensify border disputes, Congress declares war on Britain.			
1814	British soldiers invade Washington DC, and burn the capitol; Francis Scott Key writes the "Star Spangled Banner," which later becomes the US National Anthem.		*1096* – The first crusade. *1099* – Jerusalem is occupied and becomes a Christian kingdom. Several other Christian kingdoms are established in the Middle East.	
1820	Congress passes the Missouri Compromise, allowing slavery in that state.			
1821	Emma Willard founds Troy Female Seminary, the first women's college		*1187* – Saladin, Sultan of Egypt and Syria, retakes Jerusalem, giving rise to the third crusade, in which, among others, Richard the Lionheart of England takes part.	
1827	The first Black US newspaper, Freedom's Journal, is published.			
1828	The Democratic Party is established; first Native American newspaper, Cherokee Phoenix, is published.		*1192* – Japan – The Shogun (supreme commander) takes power from the emperor.	
1831	Black activist Nat Turner leads a rebellion of Virginia slaves.			
1832	Oberlin College, OH, becomes the first co-educational college.		*1196–1206* – Genghis Khan establishes a Mongol empire and begins expansion towards China.	

EUROPE	NORTH AND CENTRAL AMERICA	SOUTH AMERICA		
England 876–78 – Danish vikings create a kingdom in eastern England – the Danelaw. **900–1200** *– The Romanesque style begins to develop.; large church buildings with rounded arches and crossing vaults, decorated with moulded sculptures. Stylized painting. Medieval art is almost exclusively in the service of the Church. The Bible is the great source of inspiration. Feudalism gains inroads in Europe.* **962** *Otto the Great, German king, takes the title of Emperor of the Holy Roman Empire.* **1054** *– Investiture struggle between the emperor and the pope over control of the bishoprics.* **1054** *– Schism divides Christendom into the Roman Catholic and the Greek Orthodox Church.* **1066** *– William the Conqueror, Duke of Normandy, victorious in the Battle of Hastings, becomes king of England.* **1086** *– The Domesday Book established by William; one taxation standard for all land in England.* **1099** *– Period of wealth and power for Venice begins, based on trade with the Orient. The Church of St. Mark is built, and later the palace of the Doge. The crusades brought Europe into contact with Arab culture, which had in turn retained much of the Greek heritage.* **1088** *– University established in Bologna.* **1100s** *– The Gothic style (pointed arches) has its breakthrough. Construction begun on the cathedrals of Notre Dame de Paris and Notre Dame de Chartres. The first stern-rudders, making long sea voyages much easier.* **Ca. 1100** *– The ballad of Roland (Rollo) depicts battle at which the Franks stop the Arab advance at Roncevaux.* **Ca. 1100** *– Russia; the Nestor Chronicle depicts the founding of Russia.* **1160?** *– University est. in Paris.* **1170** *– University est. in Oxford.* **1173–1360** *– Construction begun on the Leaning Tower of Pisa.*	**800–900** *The Toltecs create an empire with an advanced culture in the Mexican highlands. Centered in Tula.* **1100s** *– Decline of the Toltec empire in Mexico.*	**Ca. 1100** *– Moche kingdom evolves into the Chimu empire, with structured social and political organization – model for the later Inca empire.*	*Shipwrights in Normandy.* *Chinese official.* *Crusader knights.* *Roman church. 1000 AD.* *Richard the Lionheart in battle.* *A Mongol and his horse.*	**800–1200**

	UNITED STATES	AFRICA	ASIA	AUSTRALIA
		The Mandingo people found the Muslim kingdom of Mali, with Timbuktu as the leading centre of trade.	Japan – *1200s* The ritual No drama is born. India ink drawing is developed.	
1836	Texans at the Alamo in San Antonio are besieged by Mexicans under the command of Santa Anna; Texas declares its independence from Mexico; the first white women cross the plains.		China – *1200s* The classical Chinese drama develops.	
1838	The US government forces Cherokees to leave their homes in the Southeast and march 1,200 miles on what was called the "Trail of Tears."			
1844	Samuel F.B. Morse sends the first telegraph message.			
1846	War with Mexico ignites over disputed Texas land.			
1848	The Mexican War ends with the US gaining vast tracts of land in the West; gold is discovered in California; Lucretia Mott and Elizabeth Cady Stanton conduct the first women's rights convention in Seneca Falls, NY.			
1850	Senator Henry Clay's Compromise admits California to the Union as a non-slave state.			
1854	The Republican Party is formed in Ripon, WI; Henry David Thoreau publishes Walden; in the Dred Scott decision, the US Supreme Court upholds slavery and denies citizenship to Blacks.			
1858	Presidential candidates Abraham Lincoln and Fredrick Douglas debate in Illinois.		*1271* – Venetian trader Marco Polo goes to China. In 1295 he returns to Venice.	
1859	Abolitionist John Brown raids a US arsenal at Harper's Ferry, WV; Brown is hanged for treason.		Japan *1274 & 1281* Mongols under Kublai Khan make failed attempts to conquer Japan.	
1860	Abraham Lincoln is elected the 16th president; the Pony Express begins service between Sacramento, CA and St. Joseph, MO.		*1279* – Genghis Khan's grandson, Kublai Khan, conquers all of China and begins the Mongol dynasty.	
1861	Seven pro-slavery Southern states cede from the Union and, under the leadership of Jefferson Davis, set up the Confederate States of America; Civil War breaks out at Fort Sumter, SC.	In the *1320s* The Muslim king of Mali, Mansa-Musa, undertakes a pilgrimage to Mecca, and upon his return begins construction of the Great Mosque in Timbuktu.		
1862	The Homestead Act promotes settlement of the Midwest by granting land to farmers.			
1863	Harriet Tubman's "Underground Railroad" ushers hundreds of slaves to freedom in the North; Lincoln issues the Emancipation Proclamation, freeing the slaves, and delivers the Gettysburg Address.			

EUROPE	NORTH AND CENTRAL AMERICA	SOUTH AMERICA		
1200s – A time of blossoming for the cities. The universities become centres of learning and culture. The writings of Aristotle are studied intensively. Romantic chivalric poetry popular among the upper classes. The many cloisters are centres of teaching, medical care, and the development of agricultural methods. Italy **1210** – Francis of Assisi founds the Franciscan order – beggar monks. England **1215** – Magna Carta; protection against arbitrary judgments by the crown in legal and taxation matters. Spain **1216** – Dominican order founded. France **1235** – Leoninus, the first known composer, dies. Choral music evolves. England **1245** – Construction begun on Westminster Abbey, which becomes the site of all future coronations. England **ca. 1259** – The English parliament begins to take form. England **1300s** – Wool trade flourishes. **1300s** – The plague (Black Death) strikes Europe and Asia. Ca. 75 million people thought to have died. Lack of manpower and changes in climate likely causes of decline in agricultural production. Transition to money-based economy, with inflated prices and devaluations. Social unrest leads to many rebellions. Secular music begins to take its place alongside church music. Italy **1303–05** – Giotto completes his frescos in Padua. He can be said to have prefigured the Renaissance through a more humanistic approach to religious motifs. Dante writes (ca. 1310–20) his Divine Comedy, in which the journey through Hell, Purgatory and Paradise reflect the Medieval world-view. In Italy the great writers, Dante, Petrarch and Boccaccio, shape the modern Italian language. **1337–1453** The Hundred Years' War between England and France. France **1309–1377** – Popes forced to reside in Avignon. The popes' «Babylonian Captivity».		**1300s** – The Incas begin building their empire on South America's west coast.	Man wheeling his wife home. Wine grapes are trampled. Mongol horseman. Ploughing with a harrow. Mistreated prisoners. Plague doctor wearing protective mask.	1 200 1 340

	UNITED STATES	AFRICA	ASIA	AUSTRALIA
			China – **1300s** The Mongol empire is overthrown and the Ming dynasty commences. Chinese renaissance. The Great Wall reaches its present dimensions. Working in porcelain (Ming porcelain) flourishes. Ming porcelain is the forerunner of the so-called East-Indian porcelain which is later exported to Europe.	
1864	Awaiting a surrender agreement, hundreds of Cheyenne and Arapahoes are massacred by US cavalrymen at Sand Creek, CO.			
1865	The Civil War ends with General Robert E. Lee surrendering at Appomattox Courthouse, VA; Lincoln is assassinated; the 13th Amendment, abolishing slavery is passed.		Persia – Hafiz (1325–1390), Persian lyric poetry's great master, writes his collection of verse – Divan.	
1866	The Ku Klux Klan is formed secretly in Pulaski, TN, to terrorize Black voters.		**1399** Mongol emperor Timur Link conquers India, Persia and Asia Minor. With his death in **1450**, the empire falls apart.	
1867	The US buys Alaska from Russia for $7.2 million.			
1869	Central Pacific and Union Pacific railroad tracks are joined at Promontory, UT uniting the East and West coasts.			
1870	Victoria Claflin Woodhull is the first woman to run for president.			
1872	Congress establishes Yellowstone as the first national park; Women's suffrage leader Susan B. Anthony is arrested for voting.			
1875	Congress passes the Civil Rights Act, giving Blacks equal access to public places and jury duty privileges. (The US Supreme Court invalidates it in 1883.)			
1876	Members of the Sioux, Northern Cheyenne and Arapahoe Nations, in part led by medicine man Sitting Bull, defeat General George Custer in the "last stand," Battle of the Little Big Horn, Montana; Custer, 264 7th Cavalry soldiers and an estimated 30 Native Americans are killed.			
1877	The US violates a land treaty with the Dakota Sioux by seizing the Black Hills in South Dakota; eleven leaders of the Molly Maguires, a society of Irish miners in Scranton, PA, are hanged for killing mining officials, and policemen.	Portugal **1432** Prince Henry («The Seafarer») organizes voyages of exploration along the west coast of Africa. The Portuguese begin construction of fortified trading stations along the west coast. The birth of a colonial empire.		
1879	Thomas Edison invents the electric light bulb; F.W. Woolworth opens his first five-and-ten cents store in Utica, NY.	**1442** – The Portuguese begin exploiting Africa's northwest coast, and find gold at Rio de Oro.		
1881	James A. Garfield, the 20th president, is assassinated soon after taking-office; Chester A. Arthur becomes the 21st president; Black educator Booker T. Washington founds the Tuskegee Institute, Alabama, for Black students.	Trade along the Atlantic coast in competition with the caravan trade across the Sahara.	Turkey – **1453** Mohammed II occupies Constantinople – the Byzantine empire ceases to exist.	
1885	The Statue of Liberty is dedicated in New York Harbor.		Russia **1453** After the fall of Constantinople, the patriarch of Moscow takes over the leadership of the Greek Orthodox Church.	

EUROPE	NORTH AND CENTRAL AMERICA	SOUTH AMERICA		

France 1358 – Peasant rebellions.

Germany 1370 – The Hansa – confederation of northern German cities – dominates trade in Northern Europe.

England 1381 – Peasant rebellions.

Poland and Lithuania enter into union in 1386.

England 1387 – Geoffrey Chaucer, «father of the English language,» publishes his Canterbury Tales.

Northern Europe 1397 – Denmark, Norway, Sweden and Finland enter into a union that exists formally until 1521.

In Northern Italy the Renaissance develops – a rebirth of Antiquity – «mankind's Golden Age». Finds expression in painting in Masaccio (1401–1428), who creates an intensively human mood in his paintings. In architecture Brunelleschi defines the style – cathedral cupola in Florence.

The search for gold and silver (to pay for the costly imports from Asia) driving force behind European explorations.

Bill of exchange becomes common form of payment. Italian system of dual-entry bookkeeping used.

1432 – Jan van Eyck completes the alter at Gent. A realistic Northern European style blossoms. With the help of a flexible new material – oil paint – it is possible to give paintings greater nuance.

France 1439 – Joan of Arc, a peasant girl from Lorraine, takes part in the Hundred Years' War to drive out the English. Taken prisoner, judged a witch and burned at the stake. Becomes a saint and national symbol.

Germany 1450 – Johann Gutenberg invents printing press with moveable type.

1 300
1 450

English wagon from the 14th century.

Arab slaves.

Timur Lenk on his throne.

Threshing grain.

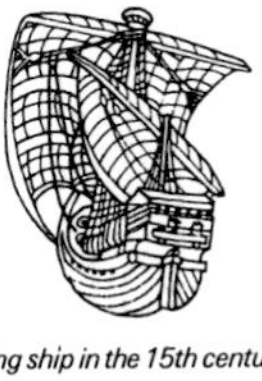

Sailing ship in the 15th century.

English archer from 1415.

	UNITED STATES	AFRICA	ASIA	AUSTRALIA
		1464–92 *Under the reign of Sonni Ali the Songhai empire reaches its largest extension in West Africa.*		
1886	*Seven policemen and four workers are killed during a bitter labor battle in Chicago's Haymarket Square; the American Federation of Labor is founded; Apache Chief Geronimo surrenders to Arizona Territory leaders.*			
1890	*Sitting Bull is killed by police officials at Standing Rock Reservation in South Dakota; an estimated 200 Sioux are massacred by US troops at Wounded Knee, SD; the Sherman Antitrust Act begins a government effort to curb monopolies; Ellis Island in New York Harbor opens as an immigration depot.*			
1896	*The US Supreme Court approves racial segregation in Plessy vs Ferguson.*			
1898	*The US blockades Cuban ports, declares war on Spain and invades Puerto Rico; Spain cedes the Philippines, Puerto Rico and Guam, and approves Cuban independence.*			
1899	*Seeking an end to US intervention, Filipino insurgents begin guerilla warfare in the Philippines; through the Open Door Policy, the US trades with China.*			
1900	*The International Ladies Garment Workers Union is founded; prohibitionist Carry Nation begins raiding Kansas saloons.*			
1901	*The US captures Emilio Aguinaldo, the leader of the Filipino insurrection, ending the movement; President McKinley is assassinated; Theodore Roosevelt becomes the 26th president; J.P. Morgan founds the US Steel Corporation.*		*China – **1498** The modern toothbrush is described in a Chinese reference work.*	
1902	*US gains control of the Panama Canal.*	**1505** – *Portuguese establish trading stations on the Zambezi river on Africa's east coast; present-day Mozambique.*		
1903	*Orville and Wilbur Wright complete the first air flight near Kitty Hawk, NC; Henry Ford establishes the Ford Motor Company.*			
1905	*Bill Haywood, Mother Mary Harris Jones, Eugene Debs, and others found the IWW– the Industrial Workers of the World union.*			
1906	*US troops occupy Cuba; San Francisco is destroyed by an earthquake.*			
1908	*Roosevelt stops Japanese immigration into the US; demanding labor improvements, women demonstrate in New York City.*			
			*India **1525** Empire of the Great Moguls is established.*	**1519–21** – *The Spanish sponsor Portuguese mariner Magellan's circumnavigation of the globe. Magellan dies during the voyage.*

EUROPE	NORTH AND CENTRAL AMERICA	SOUTH AMERICA		

EUROPE	NORTH AND CENTRAL AMERICA	SOUTH AMERICA		
Russia **1480** – Ivan III frees Moscow from the Mongol empire («Golden Horde») and joins with other princes in a Russian kingdom. Spain **1492** – The Moors (Arabs) are driven out of Spain. All Jews expelled. **1492** – Columbus discovers America, but believes he has found India. Germany **1492** – Martin Behaim constructs the first globe map. Most intellectuals now agree that the world is round – not flat. **1493** – The Spanish begin conquest of portions of North Africa. **1493** – The first Spanish settlement in the New World at Hispaniola. Italy **1495–97** – Leonardo da Vinci paints the Last Supper. **1497** – The Italian Cabot discovers Newfoundland. **1497–98** – Vasco da Gama finds the sea route to India. **1498** – Columbus discovers the South American continent. **1499** – Switzerland frees herself from the German empire. **1500s** – Beginning of the modern age. Europeans have contact by sea with Africa, Asia and the Americas. National states form. Renaissance culture has breakthrough in Europe. Church unity is broken. The printed word makes an impact in secular and religious struggles. **1516** – Erasmus of Rotterdam translates the New Testament from Greek. The Reformation Bible. Germany **1517** – Martin Luther puts forth his theses on man's way to salvation. With the printing press his ideas are spread far too quickly to suit the Church. Germany **1519** – Charles V becomes Holy Roman Emperor and inherits present-day Germany, Austria, The Netherlands, Spain, and vast areas of South America. France **1520s** – First textile factories for weaving silk. Switzerland **1525** – Zwingli's reformation in Zurich. Austria **1529** – The Turks' advance into Europe reaches it peak, and Vienna is threatened.	**1492** – Columbus discovers America. Mexico **1519–21** – Cortéz conquers Mexico and lays waste to the Aztec kingdom, the youngest of advanced cultures in the Americas. The capital of Tenochtitlàn, built on the waters of Lake Tetzcoco, is destroyed. The king, Motecuzoma, is killed. The city was at the time one of the world's largest, with some 400 000 inhabitants.	**1498** – Columbus discovers the South American continent. **1493–ca. 1527** – Huayana Capac expands the Incan empire across Ecuador to Chile.	 Europeans being carried by East Indians. Goa in India ca. 1500. Aztecs being interrogated by Cortez. Spaniard beating an Indian. Aztec woman at her loom. Holy Roman Emperor Charles V.	1475 1530

	UNITED STATES	AFRICA	ASIA	AUSTRALIA
1909	Admiral Robert E. Perry reaches the North Pole; Walter White helps found the National Association for the Advancement of Colored People (NAACP); "Free Speech Movement" springs up in cities in Washington, Montana, and California: thousands of people are arrested for speaking on street corners.			
1911	At the Triangle Shirt Waist Company in New York City, an estimated 146 textile workers, most of them women and children, die in a fire.			
1913	Income tax collection is ratified under the 13th Amendment.		Japan – **1542** The Portuguese land in Japan and begin preaching Christianity, which is forbidden.	
1915	More than 100 Americans aboard the British cruise ship Lusitania die when it is sunk by a German submarine; 25,000 women march in New York City demanding the right to vote.		**1557** – The Portuguese conquer Macao on the coast of China – remains a Portuguese possession to this day.	
1916	Due to the persistent work of Margaret Sanger, public birth control clinics open in Brooklyn, NY; the National Park Service is formed; Jeanette Rankin, Montana, becomes the first US Congresswoman.			
1917	The US declares war on Germany after German submarines wage "unrestricted" warfare; Puerto Rico becomes a US territory.			
1918	When the war ends in November, more than one million American troops are in Europe. Fifty thousand American soldiers died there.			
1919	The US Supreme Court holds that free speech doesn't apply to draft resistance (Later decisions overturn this ruling.); general strike of 100,000 shuts down Seattle, WA for five days.			
1920	Women get the right to vote under the 19th Amendment; during the "Red Scare," nearly 4,000 people are arrested and deported as alleged communist spies.			
1922	Rebecca L. Felton, Georgia, becomes the first woman US senator.			
1924	Congress recognizes Native Americans' citizenship.			
1925	John T. Scopes is fined $100 for teaching evolution in a Dayton, TN high school.			
1927	Charles Lindbergh completes the first solo flight across the Atlantic.			

EUROPE	NORTH AND CENTRAL AMERICA	SOUTH AMERICA		
England **1534** – Henry VIII breaks with the pope and founds the Church of England. Manages to outlive six wives. Henry's daughter, Mary, attempts to reintroduce Catholicism and marries Philip II of Spain. Her brutal methods gave her the nickname «Bloody Mary», perhaps best remembered in the cocktail «Bloody Mary», made from vodka and tomato juice. Germany **1530s** – The financial house of Fugger at the height of its power. Spain **1540** – The Jesuit order founded – to counter the Reformation. Switzerland **1541** – Calvin's reformation in Geneva. **1543** – Copernicus presents his view of the world, where the Sun is the centre of the universe, and Earth only a small part of the whole. Germany **1556** – Emperor Charles V abdicates. The Habsburg empire is divided. **1558–1603** – Elizabeth I reigns – the Elizabethan era–a shining epoch in the history of England. **1558** – Russian czar Ivan the Terrible advances to the Baltic. **1571** – The Spanish defeat the Turkish fleet at Lepanto. Turkish hegemony over the eastern Mediterranean ended. One of the combatants is Cervantes, who later writes Don Quixote. **1570s** – Palestrina and Orlando di Lasso develop and perfect Gregorian church music. Venice **1570** – Andrea Palladio publishes «Four Books on Architecture», which has great influence on the art of building construction. France **1572** – Religious wars (Huguenot wars) and wars of succession. Peace when Henry of Navarre made king after converting to Catholicism. «Paris is well worth a mass.» **1587** – Mary Stuart, queen of Scotland, executed for slandering Elizabeth I and making claims to the English throne. **1588** – Philip II of Spain sends the «Great Armada» against England. The English fleet, incl. circumnavigator Sir Francis Drake, victorious.	**1585** – English found the colony of Virginia on the east coast of North America.	**1531–1533** – Spaniard Pizarro's conquest of the Inca empire in Peru. Areas conquered by the Spanish organized into viceroyalties: New Spain (Mexico), the vice-royalty of Peru, and New Granada (Colombia). The Europeans in South America gained knowledge of a number of cultivated plants and vegetables: potatoes, tomatoes, maize, and tobacco.	Two duellists. Indian prince with his courtesan. A printshop in 1568. Sir Walter Raleigh.	1520 1580

	UNITED STATES	AFRICA	ASIA	AUSTRALIA
1929	Stock prices plummet, the market crashes and the Great Depression begins; much of the Midwest suffers under a drought.			
1931	The Empire State Building opens.			
1932	Charles Lindbergh's baby son is kidnapped and found dead.			
1933	Franklin Delano Roosevelt becomes the 32nd president; US Secretary of Labor Frances Perkins becomes the first woman cabinet member.		Japan – **1603** The Tokugawa period commences. After a short time of openness and expansion, Japan again closes herself to the world. A strictly controlled and hierarchical society is imposed. Continues until 1868.	
1935	Under the Wagner Labor Relations Act, workers are permitted to organize unions; Roosevelt signs the Social Security Act.			
1936	Roosevelt is re-elected; at the Olympic Games in Berlin, Black US athlete, Jesse Owens wins four gold medals, challenging Hitler's notion of Aryan racial superiority.		India – **1605** Akbar the Great dies. During his reign the empire of the Great Mogul is expanded, achieving a stabile military and political organization.	
1937	Aviator Amelia Earhart and co-pilot Fred Nooman disappear during a cross-Pacific flight.			**1606** – The Dutchman Willem Jansz sails through the Torres Strait between Australia and New Guinea and goes ashore at the eastern part of the Bight of Carpentaria.
1938	The national minimum wage is enacted.			
1940	Roosevelt takes the presidential oath for a third time.			
1941	After Japanese fighter pilots bomb Pearl Harbor, the US declares war on Japan; war is declared on Germany and Italy.			
1942	Japanese-Americans are forced into detention camps for the duration of the war. Their homes, property, and businesses are confiscated.			**1642** – Dutch seafarer Tasman sails around Australia and discovers New Zealand and Tasmania.
1944	Roosevelt is elected for a fourth time; US and other allied forces stage the D-Day invasion of Normandy.			
1945	Roosevelt dies, and Harry S. Truman becomes the 33rd president; Germany surrenders; US drops the first atomic bombs on Hiroshima and Nagasaki, and Japan surrenders.		India **1631** Construction of the queen's mausoleum Taj Mahal in Agra is begun. China – **1644** The Manchurian Qing clan, China's last imperial dynasty, takes power and a new period of greatness begins.	
1946	US Navy tests the atom bomb in the Bikini Islands.		– **1645** The Russians reach the Pacific Ocean.	
1947	Congress passes the Taft-Hartley Act to curb labor union strikes; under the Marshall Plan, the US gives war–torn European countries aid; the Central Intelligence Agency (CIA) is established; Jackie Robinson becomes the first Black major league baseball player.	**1652** – Dutch found the Cape Colony on the southern point of Africa.		

EUROPE	NORTH AND CENTRAL AMERICA	SOUTH AMERICA		

| | | | | 1600 |
| | | | | 1650 |

1600s – are marked by religious conflict and wars. North America is colonized. Birth of modern science. Emergence of the Baroque. Hansa trade monopoly is broken. England, The Netherlands and France become the new trading nations, as seen in their establishing of the East India companies. Baroque painting reflects in colour, light and movement, people's feelings during an era of unrest.

The 17th century's turmoil and struggle for power reflected in Baroque architecture. Monumental and full of life. The Baroque is spread over all of Europe and Latin America. Through the work of Galileo, Kepler and Newton, a new world-view is formed, based on mathematical formulae.

1600, Shakespeare's great tragedies appear, among others, Hamlet.

England **1617** – William Harvey discovers the circulatory system.

1618 – Thirty Years' War begins. Conflict over religious and political power in Europe.

1622 – First potatoes grown in Germany.

France **1625** – Peter Paul Rubens paints 18 panels in the Palais de Luxembourg in Paris. Baroque art is established.

Sweden **1628** – The royal ship the Vasa sinks on its maiden voyage, and isn't recovered until 1956.

Holland **1632** – Rembrandt paints «Doctor Tulp's Anatomy Lecture».

1635 – Acadmie Française founded.

1648 – Thirty Years' War ends. Germany remains religiously and politically divided. France and Sweden great-powers.

France – In the **1650s** Louis XIV begins construction of the palace of Versailles, where his court became the model for European royalty.

Establishment of absolute monarchy. «The state is me,» as Louis is supposed to have said. His Allonge wig. Symbol of power and strength, modelled on the lion's mane.

Italian Monteverdi (1567–1643) and Englishman Purcell develop opera as art form.

1662 – Royal Society founded in London.

1607 – Henry Hudson sails to Greenland, and up the Hudson River.

1608 – The French found Quebec in Canada.

1620 – The Pilgrim Fathers, radical protestants, land the «Mayflower» in Massachusetts, where they found the city of Boston in 1630.

1626 – The Dutch found New Amsterdam (later New York). The French push further inland along the St. Lawrence. Interested mostly in trading (furs), not in farming.

1640s and 50s – English royalists colonize the southern parts of the North American east coast. Plantation farming with imported black slaves.

Transport by sleigh ca. 1600.

Soldier with musket ca. 1600.

Combat during the Thirty Years' War.

Peace is proclaimed in 1648.

The Tower of London in 1641.

Prince James playing tennis.

	UNITED STATES	**AFRICA**	**ASIA**	**AUSTRALIA**
1950	Truman sends US troops to South Korea to stop invasion from North Korea.			
1951	Presidents are limited to two terms in office.			
1952	Jonas Salk develops a polio vaccine; the first hydrogen bomb is exploded over the Pacific Ocean.			
1953	The Korean War ends; Julius and Ethel Rosenberg are executed for espionage.			
1954	Senator Joseph McCarthy (R-Wis.) is reprimanded for "witch hunting" during the House Unamerican Activities Committee hearings; in Brown vs the Board of Education, the US Supreme Court rules racial segregation in schools unconstitutional.			
1955	Rosa Parks refuses to give her bus seat to a white man in Montgomery, AL, touching off the "Great Montgomery Bus Boycott"; America's two largest labor unions merge to form the AFL-CIO.	**1700s** The export of slaves from Africa is estimated at 15 million.	India – **1700** The Mogul Empire is weakened, giving the English an opportunity to control the India trade. They begin exporting Indian opium to China.	
1958	Explorer 1 becomes the first earth-orbiting satellite; the National Aeronautics and Space Administration (NASA) is established.		China – **1715** The English granted permission to establish trade mission in Canton.	
1959	Alaska and Hawaii become the 49th and 50th states.			
1960	Black College students begin staging "sit-ins" after four Black students are denied service at a Woolworth's lunch counter in Greensboro, NC.			
1961	John F. Kennedy becomes the 35th president; Alan B. Shepard, Jr. completes the first manned space flight aboard a Mercury capsule.			
1962	John Glenn orbits the earth in the space capsule Friendship 1; in spite of campus rioting, James Meredith becomes the first Black student to enroll at the University of Mississippi.			
1963	President Kennedy is assassinated in Dallas, TX; Lyndon B. Johnson becomes the 36th president; the US Supreme Court bans prayer in public schools and rules that states must provide free legal counsel for indigents; Civil Rights leader Martin Luther King, Jr. delivers his "I have A Dream Speech".		**1756–1763** – The colonial war between England and France strengthens British influence in India.	

EUROPE	NORTH AND CENTRAL AMERICA	SOUTH AMERICA		
France **1670s** – Paris becomes cultural capital of Europe. England: Declaration of Rights **1689** defines the power of the parliament and rights of individual citizens. **1690** – John Locke publishes «On Human Understanding». Due to his ideas on human rights and freedoms, he becomes the father of enlightenment and liberalism. **1694** – Bank of England established. Peter the Great begins modernizing his empire according to Western model, and expands into the Baltic. **1700** – Peter introduces a tax on facial hair, and forcibly clips the beards of noblemen. Germany **1710** – Saxons duplicate art of porcelainmaking, which has been a Chinese secret. Meissen factory founded. Germany **1710** – Leibniz formulates the problem of 'theodicy': how to reconcile the concept of a good and omnipotent God, with all the misery that afflicts the world. In Great Britain two novels published which are important to the Enlightenment: Defoe's «Robinson Crusoe» (1719) and Swift's «Gulliver's Travels» (1726). Germany Baroque music flourishes; J.S. Bach: Matteuspassion (1729) and G.F. Händel: The Messiah (1742). France **1748** – Montesquieu's On the Spirit of the Laws – source of inspiration for – among other things – the U.S. constitution, with its separation of powers. France **1759** – Voltaire's novel «Candide» published as piece of social criticism. Great Britain **1760** – Macpherson publishes «The Works of Ossian», gives impulse to Pre-Romanticism. Russia **1762** – Catherine II (the Great) regent. Her love-life presumably the most costly in history: equivalent to $1,600,000,000 (U.S.) **1762** – Rousseau: The Social Contract. Attack on private ownership as the root of all evil. England **1776** – Adam Smith: «On the Wealth of Nations» – basis of classical liberal economy.	**1755–1763** – Colonial war between England and France gives the English all land east of the Mississippi and all of Canada. **1767** – Mason-Dixon Line drawn across the eastern colonies. Separates slave-holding colonies from those which do not accept slavery.	**1713** – Importation of slaves from Africa to the Spanish colonies. The English slave trade reaches its peak.	Harvesting in 1675. English pillory. 1700s. Apartment buildings in Paris. 1700 s. French execution in 1757. Basic steps in the Minuet.	**1690** **1770**

	UNITED STATES	AFRICA	ASIA	AUSTRALIA
1965	Anti-war protests escalate due to US military intervention in Vietnam; on the third try, Martin Luther King, Jr. leads civil right marchers from Selma to Montgomery, AL. This attempt is successful because of federal mobilization of National Guard Units to protect the marchers.	Egypt – **1798–1799** Napoleon's expedition to Egypt makes the country a sought-after area for archaeological research. Among other things, the so-called Rosetta Stone is found, which leads to an interpretation of the hieroglyphs (1822).		**1770**–James Cook lands in Australia and claims the land for the British crown.
1966	The National Organization for Women (NOW) is founded.	The British fleet under Nelson defeats the French at Abukir.		
1967	66 die in Newark, NJ, and Detroit, MI race riots; Thurgood Marshall is sworn in as the first Black US Supreme Court justice.	West Africa – **1805** The Fulani people occupy the Hussa kingdom and establish a Muslim empire with Sokoto as its capital. The beginnings of present-day Nigeria.		
1968	Martin Luther King, Jr. is assassinated in Memphis, TN; Senator Robert F. Kennedy is assassinated in Los Angeles; Representative Shirley Chisholm, New York, is the first Black woman elected to Congress; a coalition of women's groups disrupt the Miss America Pageant in the first mass demonstration of the modern day women's movement.	Egypt – **1811** Mohammed Ali occupies the country, which during his reign becomes more independent of the Turks.		**1788**–Botany Bay becomes a penal colony. **1793**–The first voluntary immigrants come ashore. **1797**– Lieutenant John MacArthur introduces the Spanish Merino sheep, laying the cornerstone for future sheep raising.
1969	Richard M. Nixon becomes the 37th president; Neil Armstrong, becomes the first man to walk on the moon; the modern day gay rights movement starts with the Stonewall rebellion in a New York City bar.	Southeast Africa – **1816** The Great founds a Zulu empire, which continues until 1887. West Africa – **1822** Liberated black slaves found the state of Liberia.	India – **1818** The British East India Company now controls the entire country. The Malacca peninsula **1819** sir Stamford Raffles founds an English colony on the southern tip of the peninsula – Singapore. Due to its location and free trade, the city quickly becomes an important trading centre.	**1802–03** – Matthew Flinders sails around the entire continent and charts the coastline. He proposes to name it Australia.
1970	National Guardsmen kill students at Kent State University, OH, and Jackson State, MS during anti-war demonstrations; members of the "Chicago 7" are acquitted of conspiring to incite a riot during the 1968 Democratic National Convention.	North Africa – **1830** The French begin the conquest of Algeria, which later comes under French administration. Southern Africa **1836** The Cape Colony's original colonizers, the Dutch-known as the Boers-, dissatisfied with British rule, migrate northward (The Great Trek) and found Transvaal and the Orange Free State.	China – **1839–42** China halts the import of opium, leading to war with England – the Opium War. The British take Hong Kong.	
1971	The 26th Amendment gives 18-year-olds the right to vote; Lieutenant William L. Calley, Jr. is convicted of killing Vietnamese civilians, many of them women and children, in the 1968 My Lai massacre.	**1840** – The Boers defeat the Zulus in the Battle of Blood River.	**1850** The Taiping Rebellion – history's most destructive and bloody civil war; ca. 17 million dead.	**1840** – The deportation of prisoners to Australia is halted. **1850** – Australia's first university founded in Sidney.
1973	In Roe vs Wade, the US Supreme Court, in part, bans state anti-abortion laws; Nixon refuses to release tapes relevant to the Watergate break-in; two of the Watergate "burglars" are convicted; in the "Saturday Night Massacre," Attorney General Elliot Richardson resigns and Nixon fires special prosecutor Archibald Cox and William Ruckelshaus when they try to secure a court order forcing him to release Watergate tapes.	**1853** – The Scottish missionary Livingstone begins his explorations of Africa.	India – **1853** The country gets its first railways and telegraph lines. Japan – **1853** An American naval squadron under commodore Perry forces Japan to open itself to the West, leading to the demise of the shogunate and the fedual system. The rapid industrialization of Japan begins.	**1851** – Gold-rush to Bathurs and Ballarat. **1854** – Gold miners and the military clash in Eureka – the only armed conflict on Australian soil.
1974	Under public pressure over the "Watergate Scandal," President Nixon resigns; Vice President Gerald Ford becomes the 38th president, pardoning Nixon for all federal crimes he "committed or may have committed."		India – **1857** Sepoy Rebellion breaks out, led by the native soldiers of the Company's army – the Sepoys. The East India Company loses its control of India, which is placed under the British crown.	

EUROPE	NORTH AND CENTRAL AMERICA	SOUTH AMERICA		
Germany **1781** – Immanuel Kant publishes «Critique of Pure Reason,» where he discusses the limits of human knowledge. One of Romanticism's philosophical starting points. France **1783** – Flight of the first manned balloon, constructed by the brothers Montgolfier. England **1780s** – The Industrial Revolution is in full swing. Seen in the textile industry in inventions such as mechanical spinning wheels and looms (1785); in the iron industry, the Puddel process, by which malleable iron could be produced through the addition of carbon (steel). James Watt's steam engine (1769) is improved and becomes the new power source. Austria **1780s and 90s** – Mozart's great works produced. France **14 July 1789** – The storming of the Bastille. The French Revolution begins. Feudalism disappears. **1792** – France becomes a republic. **1799** – Napoleon seizes power. Poland **1795** – The 3rd division of Poland. In effect, Poland ceases to exist. Austria **1800–1810** – Beethoven creates his best known works. **1812** – Napoleon attacks Russia and occupies Moscow. The French withdrawal in the bitter cold is a catastrophe for the French army. **1815** – Napoleon defeated by Wellington at Waterloo and sent into exile on St. Helena. France **1830** – The July Revolution inspires revolutionaries in Italy and Germany. England **1830s** – Charles Dickens, writes The Pickwick Papers and Oliver Twist. Ireland **1845** – The Great Potato Famine. Mass immigration to the USA. France **1848** – The February Revolution. France becomes a republic. Marx and Engels writes the «Communist Manifesto». **1852** – Napoleon III takes power. **1853–56** – Crimean War. Horrific conditions for the wounded inspires Florence Nightingale's epochmaking work. In 1860 she starts a school of nursing at St. Thomas hospital in London.	**1773** – Boston Tea Party. Colonists disguised as indians dump a ship's cargo of tea overboard. Bitterness due to tax on tea. **1776** – The American colonies declare their independence. «We hold these truths to be selfevident, that all men are created equal, endowed by their creator with certain inalienable rights, among which life, liberty and the pursuit of happiness.» **1783** – Peace. England recognizes the USA. **1789** – The American Constitution completed. George Washington becomes the country's first president. **1803** – France sells Louisiana west of the Mississippi to the U.S. **1804–1806** – Lewis and Clark explore the American Northwest and reach the Pacific. **1818** – The 49th parallel becomes boundary between U.S. and Canada. **1823** – Monroe Doctrine declared to prevent European intervention in American (esp. S. American) affairs. **1830** – First railroads in the U.S. **1844** – Dentist Horace Wells begins using aesthetics on his patients. **1848** – Gold discovered in California. Goldrush to the west coast – «He was a miner, forty-niner...» **1852** – Henry Wells starts the Wells Fargo delivery service to the West.			1770 1860
				Neo-classicism in Munich.
				Russian village ca. 1800.
		1810–1826 – The Spanish and Portuguese colonies become independent nations. **1822** – Brazil becomes empire – lasting until 1889. The poorly defined borders in the South American interior lead to numerous border-clashes in the 1840s and 50s.		Napoleon returns home from Russia in 1812.
				Indian locomotive ca. 1850.
				Florence Nightingale in the Crimea.

	UNITED STATES	AFRICA	ASIA	AUSTRALIA
		Egypt– **1869** The Suez Canal completed. Built by Ferdinand de Lesseps and financed by France. Verdi was inspired to write his opera Aida, set in ancient Egypt.		**1860** – Burke and Wells cross the continent from south to north, but starve to death on the return trip.
1975	US troops leave Vietnam.			
1976	July 4th, marks the country's 200th anniversary; Viking I and II land on Mars; 29 people attending a Philadelphia, PA convention die from "legionnaires' disease."	**1871** – During his search for the source of the Nile, Livingstone disappears and is later found by the American journalist Stanley.		
1978	President Carter facilitates peace treaty between Eygpt and Israel.	**1875**– To secure British access to India, prime minister Disraeli buys majority shares in the Suez Canal.		
1979	Emergency crews prevent a total melt down at the Three Mile Island nuclear reactor near Middletown, PA.	Congo – **1885** Leopold II of Belgium gains control of the area and enforces merciless	**1877** – Queen Victoria is proclaimed empress of India.	**1880** – Highwayman Ned Kelly is taken prisoner and hanged.
1980	Carter places a grain embargo against the Soviet Union for its invasion of Afghani-stan; to further protest the invasion, the US Olympic Committee decides to boycott the Moscow games; Washington State's Mount St. Helens volcano erupts, killing at least 60 people.	policies of forced labour and physical torture. This inspires Joseph Conrad to write his masterpiece, Heart of Darkness. **1886** – Gold found in Transvaal, leading to a gold-rush.	**1900** China – The Boxer Rebellion; Europeans are persecuted, and the embassies in Peking are under seige. **1904–1905** – The Russo-Japanese War. Due to	**1901** – The first parliament assembles in Melbourne. The six colonies form the Commonwealth of Australia.
1981	Sandra Day O'Connor is appointed the first woman US Supreme Court justice.	**1890** – Cecil Rhodes, who amassed a fortune from diamond mines, founds	Japan's rapid industrial and technological advancement, her fleet was able to quickly crush	
1982	Unemployment hits 10.8 percent, the highest rate since 1940; the Equal Rights Amendment is defeated; Barney B. Clark receives the first artificial heart.	Rhodesia. **1899** – Boer War between England and the Boers, who are defeated. The Boer states are incorporated in the British	the Russians. The first time in history that an Asian nation had defeated a European foe. China **1911–12** – Revolution –	
1983	AIDS, a mysterious and deadly new virus is first diagnosed in homosexual males; US Marines invade Grenada; Space Shuttle Challenger crew member Sally Ride becomes the first American woman in space.	empire. **1910** – The South African Union is formed. Member of the British Commonwealth. **1911** – Italy conquers Libya. **1922** – Following a nationalist rebellion, Egypt is liberated from	the emperor is overthrown, a republic established. Yuan Shih-kai China's first president. **1917** – The Balfour Declaration promises the Jews a new homeland in Palestine.	**1917** – Railway spanning the continent completed. 1051 miles long. **1927** – Canberra becomes seat of the Parliament.
1984	CIA officials acknowledge they helped mine Nicaraguan harbors; Geraldine Ferraro is first female vice presidential candidate; Ronald Reagan is re-elected for a second term; Dr. Katheryn D. Sullivan is first female astronaut to walk in space.	British and French influence. **1934** – Italy, led by the fascist dictator Benito Mussolini, attacks Ethiopia, Africa's oldest independent nation. UN sanctions have no effect. Ethiopia falls.	**1919** – Gandhi introduces passive resistance to British rule. **1932** – The kingdom of Saudi Arabia is formed.	
1986	First official observance of Martin Luther King, Jr.'s Birthday; the Space Shuttle Challenger explodes, killing all seven crew members, including teacher Christa McAuliffe; spacecraft Voyager II at 2,000 million miles from earth sends pictures of Uranus.		**1931–1932** Japan's expansion on the mainland continues with the conquest of Manchuria and attacks on China proper (1937).	
1987	Lt. Col. Oliver North and other key Reagan officials are questioned about their involvement in the Iran-Contra affair, in which weapons were traded for US hostages; Reagan and Russian Premier Gorbachev sign an agreement to dismantle intermediate range missiles, after a record high on Wall Street, the Dow drops 508 points in one day.			

EUROPE	NORTH AND CENTRAL AMERICA	SOUTH AMERICA		
Between the years 1845–1895 many of the things now part of everyday life are invented: the sewing machine, the refrigerator, the telephone, film, the automobile and the gramophone. Sweden **1867** – Alfred Nobel invents dynamite, his wills grants annual prizes for literature, medicine, physics chemistry and peace. **1870s** – Rapid industrialization of the U.S. initiated. The Netherlands, England and Russia. **1871** – In connection with the Franco-German War, the new, united German Empire is proclaimed at Versailles. Scandinavia **1880s** – Dramatists Ibsen and Strindberg produce their epoch-making dramas. France **1894** – Captain Alfred Dreyfus is deported to Devil Island, sentenced for treason. **1861** – Italy united. **1861** – Serfdom abolished in Russia. France **1863** – Edouard Manet, pioneer of French Impressionism, creates scandal when he exhibits his «Luncheon on the Grass». Emile Zola, writes his naturalistic novel, first great work, «Thérèse Raquin» 1867. **1911** – The Norwegian explorer Roald Amundsen reaches the South Pole. **1914** – The gunshot in Sarajevo – the Austrian heir apparent is murdered by a student, Gavrilo Pricip. The murder triggers the First World War. Russia **1917** – Revolution. The czar abdicates. The communists under Lenin take power. The first socialist state. **1918** – First World War ends. **1919** – The Peace of Versailles. Germany forced to pay reparations and greatly reduce its armed forces. New, independent states formed: Estonia, Latvia, Lithuania, Hungary, Poland, Czechoslovakia and Yugoslavia. Germany **1925** – Inflation runs wild. DM becomes worthless. Germany **1933** – Hitler comes to power.	**1860** – The Southern States secede from the union and civil war breaks out. Abraham Lincoln elected president. **1865** – The Southern States defeated. Slavery is forbidden. Lincoln is assassinated. **1869** – The Pacific Railroad completed. **1870s** – Rapid industrialization of the U.S. initiated. **1876** – Battle of the Little Big Horn. Sioux indians, led by Sitting Bull, defeat general Custer's cavalry. One of the last attempts by natives to stop the white man's exploitation of the Wild West. **1898** – War with Spain. The U.S. annexes Puerto Rico and the Philippines. At the turn of the century, jazz is born in New Orleans and entertainment district of Storyville is closed. During WW1 jazz is spread across the U.S., esp. to Chicago and New York. USA **1903** – The Wright brothers make the first motorized flight. **1909** – Henry Ford begins mass-production of the Model-T Ford. USA **1912** – Hollywood becomes centre of the film industry. **1927** – The first films 'The Jazz Singer' with sound – «talkies». **1927** – Charles Lindbergh makes the first trans-Atlantic flight. **1928** – October. The great stock market crash on Wall Street. The Great Depression begins. USA **1931** – Empire State Building. 449 meters, 102 stories high. USA **1935** – Hitchcock has breakthrough with «The Thirty-nine Steps». USA **1936** – Chaplain creates his parody of modern industrial society – «Modern Times».	 **1930** – Military coup in Brazil. **1932** – War between Bolivia and Paraguay over possession of the Gran Chaco region.	 Garibaldi helps Victor Emanuel put on the «Italian boot». American paddle-wheeler ca. 1860. Stanley and Livingstone meet. The Wright brothers fly in 1903. Amundsen at the South Pole in 1911. British tank 1914–18.	1850 1940

	UNITED STATES	AFRICA	ASIA	AUSTRALIA
1988	Nearly four million acres of forest land are destroyed by fire during the worst drought in 50 years.	**1941** – German-Italian force under Field Marshal Rommel attacks Egypt. **1943** – War in North Africa ends. General Montgomery drives Rommel's army out.	**1945** – The U.S. drops atomic bombs on Hiroshima and Nagasaki - Japan capitulates. **1947** – British India is divided into two independent states - Islamic Pakistan and Hindu India.	**1972** – The conservative Liberal Party is defeated in general elections after 23 years as the governing party. A new government is formed by the Labor Party.
1989	George Bush becomes the 41st president; Former President Reagan and Bush are subpoenaed during early Iran-Contra trials; the US Supreme Court's "Webster" ruling permits states to regulate abortions again; Lt. Col. Oliver North is convicted of destroying National Security Council documents in the Iran-Contra affair; San Francisco is rocked by a large earthquake, collapsing the Oakland-San Francisco Bay bridge.	**1949** – South Africa introduces "apartheid" - a system of laws for separation of racial groups. **1956** – Egypt's president Nasser nationalizes the Suez Canal. British, French and Israeli troops attack but have to withdraw after UN intervention. **1960–64** – The largest wave of decolonization in Africa takes place -25 independent African nations are born. **1960** – Sharpeville massacre in South Africa heightens tensions between white minority and black majority. **1961** – South Africa leaves the British Commonwealth. **1965** – Rhodesia declares	**1949** – Mao Tse-Tung's communist armies drive the nationalists to Taiwan. **1955** – The "French" part of the Vietnam war. **1966-69** –China - Mao leads the Cultural Revolution against "bourgeois tendencies". **1971** – The war between India and Pakistan results in East Pakistan separating and becoming the state of Bangladesh. **1973** – The oil-producing Arab nations cut back production and quadruple oil prices, resulting in poor economic times throughout the world. **1976** – North and South	**1973** – The renowned Opera House in Sydney completed. **1974** – Prime minister Gough Whitlam removes restrictions on non-white immigrants. **1988** – Bicentennial of the arrival of the first white immigrants. Celebrations take place in the presence of Queen Elizabeth, who dedicates new parliament building in Canberra. **1993** – The Labor Party wins general elections and remains the governing party. Labor has been in power from 1972-75 and from 1983.
1990	In response to the invasion of Kuwait, President Bush deploys the largest number of troops since the Vietnam War to the Persian Gulf.	independence under white minority government. **1975** – Former Portuguese colonies Mozambique and Angola gain independence - start of civil wars in both countries. **1980** – Rhodesia becomes independent state of Zimbabwe when the white minority gives	Vietnam unite into one country. **1976** – Chinese leader Mao Tse-tung dies. Deng Xiaoping gradually appears as the new "strong man" in China. Deng introduces a series of reforms to liberalize the economy, but without giving up the	
1991	The U.S. leads a military UN-coalition in a war against Iraq, leading to the liberation of Kuwait. President Bush's popularity among Americans reaches an all-time high, according to opinion polls. But soon afterward his popularity declines due to economic crisis and growing unemployment at home.	up power to a black government. **1985** – South Africa - Racial unrest grows and leads to a declaration of martial law (1986). **1990** – Namibia, the last colonial territory in Africa, becomes a free country after a long war for independence from South Africa. Nelson Mandela, leader of African Liberation Organization (ANC), is freed after decades of imprisonment. President F. W. de Klerk orders martial law abolished.	communist party's monopoly on political power. **1979** – The Shah of Iran is overthrown and Iran becomes an Islamic republic, led by the fundamentalist Shiite, Ayatollah Khomeini. **1979** — Soviet troops invade Afghanistan. **1980** – The bloody Iran-Iraq war begins. Peace 1988. **1982** – Israel invades Lebanon. The Palestine Liberation Organization (PLO) is expelled from Beirut.	
1992	More than 40 people killed in race riots in Los Angeles. The riots start when a white jury gives the verdict of not guilty in a trial of four white policemen accused of having used excessive force while arresting, Rodney King, an African American.	**1991** – South Africa - Several basic apartheid laws abolished. Start of CODESA (Conference for Democracy in South Africa) - negotiations between black and white leaders for new democratic constitution and black franchise. Civil war in Angola ends with peace agreement. Communist dictatorship in Ethiopia is	**1984** – The Bhopal catastrophe in India: 2500 people die, as poisonous gas is leaked from a Union Carbide factory. **1988** – The Soviet Union leaves Afghanistan. **1989** – China - The communist dictatorship crushes a student movement for democracy with military force, killing hundreds of student demonstrators in a	
1992	George Bush defeated in the presidential elections by the Democratic candidate Bill Clinton.	overthrown after bloody civil war. Victorious guerilla forces form a new government with promises to lead the country toward democracy. **1992** – South Africa - Massacre	massacre in Beijing. **1990** – Iraq invades Kuwait. **1991** – Alliance of more than 30 states, led by the U.S., defeats Iraq and liberates Kuwait in a	
1993	Bill Clinton becomes the 42nd president of the United States, bringing the Democrats back to the White House after twelve years of Republicans Ronald Reagan and George Bush.	in black township of Boipathong signifies new peak of violence between ANC and black rival group Inkatha. ANC accuses government of encouraging black-on-black violence. CODESA negotiations collapse for the time being. New outbreak of civil war in Angola when guerilla faction, Unita, refuses to accept result of first democratic elections. Peace agreement ends civil war in Mozambique. Disastrous famine in Somalia due to combination of civil war and drought. Intervention of US troops (under UN supervision) to secure distribution of food. **1993** – South Africa - One of the most important ANC leaders, Chris Hani, murdered by white extremists. New threat against peace talks between black and white South Africa.	war lasting six weeks. The decision to go to war is based on a resolution taken by the UN security council. **1991** – Breakup of the Soviet Union leads to creation of five new independent states in (formerly Soviet) Central Asia. **1991** – Peace conference starts on the Middle East (with inaugural meeting in Madrid). Israel and its Arab adversaries agree for the first time to meet eye to eye at the same conference table. **1992** – Afghanistan - The communist dictatorship is overthrown. The combined forces of "mujaheddin" (islamic guerilla groups) take power in Kabul. Civil war continues, however, due to new fighting between rival guerilla groups. **1992** - India - Bloody fighting between Hindus and Muslims takes hundreds of lives. The cause of the violence is a Muslim mosque destroyed by a mob of Hindus.	

EUROPE	NORTH AND CENTRAL AMERICA	SOUTH AMERICA		

EUROPE

1939 – Hitler invades Poland.
1945 – Second World War ends. Germany defeated. 40 million people (of which 6 million Jews) have died.
1949 – North Atlantic Treaty signed (NATO).
1953 – Joseph Stalin dies.
1955 – The Warsaw Pact formed. The Cold War develops.
1956 – Hungary - Popular uprising against the communist regime. Soviet troops called in to quell the revolt.
1961 – The Berlin Wall is erected. The USSR sends up the first manned spacecraft.
1964 – Soviet Union - Khrushchev toppled when his economic policies fail.
1968 – Czechoslovakia - Attempt made to liberalize the communist system - known as the "Prague Spring" - put down by Soviet troops.
1974 – Portugal becomes democracy.
1975 – Spanish dictator, General Franco, dies. Democracy begins.
1980 – Poland - The trade union, Solidarity, is formed and begins the struggle for democracy.
1985 – Gorbachev comes to power in the Soviet Union.
1986 – Melt-down at the nuclear reactor in Chernobyl, Soviet Union.
1989 – Democratic elections in Poland. Gorbachev launches "perestrojka". Romanian dictator Ceausescu executed. Dismantling of the Berlin Wall begins. A wave of liberation sweeps over Eastern Europe.
1990 - Germany reborn as one united federal republic with the unification of East and West Germany. Democratic elections take place for the first time in 40 years in several former communist countries in East Europe. The Warsaw Pact (communist military alliance) is dissolved.
1991 – Soviet Union disintegrates and is dissolved. The 15 former Soviet republics emerge as independent states. EEC-countries of western Europe sign a treaty in Maastricht to develop a tighter economic and political union. Yugoslavia breaks-up - former Yugoslav republics Slovenia, Croatia and Macedonia issue declarations of independence. War starts in former Yugoslavia with fighting between Serbs and Croats in Croatia.
1992 — Cease fire and arrival of UN troops establishes a fragile peace in Croatia. New war starts between Serbs, Croats and Muslims in Bosnia-Herzegovina. EEC is plunged into s severe crisis. Denmark votes against the Maastricht treaty in a referendum. Opposition to Maastricht growing elsewhere in the Community.
1993 – Czechoslovakia splits up and becomes two countries - the Czech Republic and Slovakia. War in Bosnia continues, despite mediating efforts by UN and EEC.

NORTH AND CENTRAL AMERICA

1941 – Japan's attack on the American fleet at Pearl Harbor in Hawaii leads to U.S. entering the Second World War.
1944 – USA - The first computer is built greatest development comes after 1971 when the microchip is developed.
1945 – The United Nations (UN) is formed and headquartered in New York.
1954 – Hemingway receives Nobel Prize for literature as leader of the "hard-boiled school".
1962 – UN - Krushchev pounds his shoe on the podium when displeased with General Secreraty Hammarskjöld.
1962 – USA - Filmstar Marilyn Monroe dies of drug overdose.
1963 – President Kennedy murdered in Dallas.
1964 – Laws passed against racial discrimination.
1968 – Black civil rights activist Martin Luther King is murdered.
1971 – President Nixon and Secretary of State Kissinger initiate detente with China and the Soviet Union.
1974 – President Nixon forced to resign over the Watergate Scandal.
1979 – Civil war in Nicaragua. Sandinistas come to power.
1981 – Ronald Reagan becomes president in U.S.
1986 – Sparacraft Voyager II sends back pictures of the planet Uranus from a distance of 2 billion miles.
1987 – Reagan and Gorbachev reach agreement in Washington on banning land-based medium range ballistic missiles.
1989 – Panama-Dictator Manuel Noriega is deposed through military invasion by U.S. troops. Noriega is brought to the U.S. to stand trial for complicity in cocaine smuggling.
1990 – Democracy is born in Nicaragua. The communist Sandinistas lose first free parliamentary elections. New right-center government formed and peace is proclaimed, ending the civil war in the country. Sandinistas remain influential, with a strong position in trade unions, the armed forces and other organizations.
1991 – Haiti - The country's first democratically elected president, Jean-Bertrand Aristide, is installed. President is deposed after only seven months by a military coup.
1992 - El Salvador - UN-mediated peace agreement is signed between the government of president Christiani and the leftist guerilla group FMLN. The treaty ends 12 years of bloody civil war.

SOUTH AMERICA

1946 – Argentina - Juan Perón comes to power.
1959 – Fidel Castro leads the revolution in Cuba. He introduces communism and allies the country with the Soviet Union.
1962 – Cuban Missile Crisis; the Soviet Union attempts to station long-range missiles in Cuba. Solution negotiated by the U.S. and Soviet Union via 'hot line' phone.
1970 – Salvador Allende is the first democratically elected Marxist leader in Chile.
1973 – Allende toppled in military coup lead by general Pinochet, who becomes country's dictator.
1980 – First free elections in 17 years held in Peru.
1982 – Argentina invades the Falklands Islands, but gives up following a British counter-attack.
1983 – Argentina returns to democracy after seven years of military dictatorship. Raul Alfonsin becomes president after free elections and forms a civilian government.
1985 – Democracy reintroduced in Brazil and Uruguay.
1989 – Chile - Democracy is restored after 16 years of military dictatorship (led by Augusto Pinochet). The Christian Democratic candidate Patricio Aylwyn wins the first free election in 20 years.
1989 – Paraguay - Alfredo Stroessner, veteran among South American dictators, deposed in a military coup. The coup leader, A Rodriguez becomes the new president in elections where opposition is not given equal opportunities.
1992 – Peru - Democracy suspended in a military coup conducted by the democratically elected president, Alberto Fujimori. He defends the action as necessary to save democracy in the long run and also necessary to give the country strong leadership in fighting the leftist guerilla groups.
1992 – Brazil - President Fernando Collor de Mello is found guilty of corruption and is suspended from office to be impeached. He resigns before the end of the year.

American aircraft carrier.

UN'insignia from 1945.

Stalin lying in state.

The Soviet Vostok from 1962.

American bomber over Vietnam.

Line of jobless people.

1939
1990

PICTURE CREDITS

In addition to the artwork collected from J.W. Cappelen's archives, the following picture sources are used:

AAA photo, Paris, p. 35 (above).
Biblioteca Angelica, Roma, p. 49 (below).
Biblioteca Riccardiana, Firenze, p. 49 (above).
Bibliothèque National, Paris, p. 41, 56 (below).
Bridgeman Art Library, London, p. 52 (to the right), 61 (below), 66, 67, 74 (below to the left), 76 (below), 81, 90 (above), 92 (below), 98 (both), 99, 104, 105 (both), 107, 110.
British Museum, London. p. 60.
Camera Press, London, p. 119 (below), 122 (all three to the right), 123 (above to the left), 126 (to the right).
Cinevision, Stockholm, p. 20.
Douglas Dickens, London, p. 39 (below).
C.M. Dixon (Photoresources), Kingston, Kent, p. 7, 9 (above), 13, 17, 19 (above), 22, 23, 25 (below), 43, 72, (above).
E.N.I.T., Rome, p. 27 (above).
The Eremitage Museum, Leningrad, p. 89 (above).
The Frederiksborg Museum, Hillerød, p. 69 (above).
Galleria Villa Borghese, Rome, p. 206.

Mats Halling, Stockholm, p. 31 (below).
Robert Harding Picture Library, London, p. 11 (above to the right and below), 14 (above), 21 (in the middle), 33 (below), 62 (above), 64 (above), 68, 71, 73 (below), 74 (above), 75 (above), 77 (all three), 85 (above), 90 (below), 96 (below).
Herakleion Museum, Athens, p. 8 (to the left).
Michael Holford, London, p. 12 (above), 15, 18 (above and below to the left), 21 (above), 25 (above), 27 (below), 28, 31 (above), 35 (below), 63 (below), 65, 89 (below and to the right), 91.
Hulton Picture Library, London, p. 54 (in the middle), 106 (below), 126 (to the left).
Robert Hunt Library, London, p. 102, 111 (above).
Imperial War Museum, London, p. 112 (below).
Iraqi Museum, Baghdad, p. 11 (above to the right).
Heikki Kirkinen, Joensuu, p. 51.
Library of Congress, Washington D.C., p. 94 (below), 109 (in the middle to the left).
Louvre, Paris, p. 19 (below), 29 (above), 73 (above).
Mansell Collection, London, p. 54 (above), 64 (below), 83 (above), 101, 106 (above).

Musée Carnavalet, Paris, p. 78 (above).
Musée Guimet, Paris, p. 39 (above).
Museo Curco, Como, p. 61 (above).
Museo del Risorgimento, Milan, p. 87 (above to the right).
Museo Nacional, Lima, p. 94 (to the left).
Museo Nacional de Antropologia, Mexico City, p. 62 (below).
National Army Museum, London, p. 97 (above).
National Gallery, p. 88 (above).
Heikki Partanen, Helsinki, p. 10, 14 (below).
Politikens Forlag, Copenhagen, p. 12 (below), 16, 18 (below to the right), 20 (above), 21 (below), 29 (below), 36 (below), 38 (below), 47 (below), 55, 61, (in the middle), 63 (above), 74 (below), 83 (below), 87 (above to the left and below), 92 (above), 103, 105 (to the left), 120 (to the right).
Politikens Pressefoto, Copenhagen, p. 108, 118 (to the right), 120 (to the left), 121, 122 (farthest left), 123 (above to the right), 127, 128, (below to the left and the two above to the right).
Popperphoto, London, p. 124 (above).
Prähistorisches Museum, Hallstatt, p. 8 (to the right).
Rex Features Ltd., London, p. 115

(below to the right), 118 (to the left), 123 (below).
Scan-Foto, Frankfurt p. 128 (below to the right).
Schloss Charlottenhof, Sans-souci, Berlin, p. 95.
Staatsarchiv, Hamburg, p. 56 (above).
Stadtbibliothek, Nuremberg, p. 58 (below).
Sundahl Foto, Stockholm, p. 69 (below).
Toledo Museum of Art, Ohio, p. 59 (above).
Universitätsbibliothek, Geneva, p. 59 (in the middle).
The Vatican Museum, Rome, p. 5 (below).
Zefa, London, p. 38 (above).

Illustrations:
Sv. Aa. Voigt Andersen, Copenhagen, p. 20 (below).
Arne Gaarn Bak, Copenhagen, p. 116.
Lars Tangedal, Copenhagen, p. 9 (below).

Whilst every effort has been made to trace copyright holders, the publisher will be pleased to make the necessary arrangements at the first opportunity if they have overlooked any.